The Advances in Fluid Mechanics

The Advances in Fluid Mechanics

Editor

Jesús María Blanco

MDPI

Basel • Beijing • Wuhan • Barcelona • Belgrade • Novi Sad • Cluj • Manchester

Editor
Jesús María Blanco
Energy Engineering
University of the Basque Country
Bilbao
Spain

Editorial Office
MDPI
St. Alban-Anlage 66
4052 Basel, Switzerland

This is a reprint of articles from the Special Issue published online in the open access journal *Applied Sciences* (ISSN 2076-3417) (available at: www.mdpi.com/journal/applsci/special_issues/Advances_in_Fluid_Mechanics).

For citation purposes, cite each article independently as indicated on the article page online and as indicated below:

Lastname, A.A.; Lastname, B.B. Article Title. *Journal Name* **Year**, *Volume Number*, Page Range.

ISBN 978-3-0365-9033-2 (Hbk)
ISBN 978-3-0365-9032-5 (PDF)
doi.org/10.3390/books978-3-0365-9032-5

Contents

About the Editor

Jesús María Blanco

Dr. Blanco has been a Doctor in Industrial Engineering since 1996 (Public University of Navarre) and a Master in Offshore Research Tidal Turbines since 2009 (Cranfield University, UK). He has also been an Associate Professor at the University of the Basque Country (UPV/EHU), School of Engineering, Department of Energy Engineering, since 1993, as well as a Tutor at the Open University (Spain) since 1997. He has been a visiting professor at Cranfield University (UK) since 2009, carrying out several research stays since 2008, funded by the Spanish Ministry of Education.

He has supervised six doctoral students and is currently supervising six other doctoral students in this field, having been a member of doctoral thesis tribunals at different national and international levels. He has published 60 JCR papers and more than 70 international conferences focused on energy optimization, CFD, marine renewable energies, and green transitions, as well as 30 other conferences focused on new trends in higher education. He has been the co-author of 2 patents (1998 and 2019), 8 teaching books with ISBN, as well as 28 books and book chapters in prestigious international publishers. He has led numerous research projects and contracts in the field of energy and sustainability, highlighting the recently finished RENOVABLES project from the Euro Region. He has been the principal investigator of the consolidated research group of the Basque Government (GIU19-029) and the recently awarded research group of the Basque Government (2021-2025) IT1415-22.

He has been the chairman of the International Congress (CMN'2013), a member of the scientific committee of several international congresses, the chairman of HPCC 2022, and the chairman of the European Wave and Tidal Conference (EWTEC 2023).

He has also been the director of the MORE (Master Offshore Renewable Energy) project since 2016, the coordinator of the Erasmus Mundus REM (Renewable Energy in the Marine Environment) project since 2018.

Preface

Fluid mechanics, especially fluid dynamics, is an active field of research, typically mathematically complex. Many problems are partly or wholly unsolved and are best addressed by numerical methods, typically using computers. It is a subdiscipline of fluid mechanics that deals with fluid flow, i.e., the science of liquids and gases in motion. Fluid dynamics offers a systematic structure—which underlies these practical disciplines—that embraces empirical and semi-empirical laws derived from flow measurement and used to solve practical problems.

Jesús María Blanco
Editor

applied sciences

MDPI

Editorial

Special Issue on the Advances in Fluid Mechanics

Jesús M. Blanco

Energy Engineering Department, School of Engineering, University of the Basque Country (UPV/EHU), Plaza Ingeniero Torres Quevedo, Building 1, 48013 Bilbao, Spain; jesusmaria.blanco@ehu.eus

The progressive implementation of computational fluid dynamics (CFD) has experienced a great increase over the last decades as its use has become more feasible worldwide. The goal of turbulence modeling is to reproduce these flow physics as accurately as possible with a reasonable computational effort. In some cases, turbulence is modeled by the Reynolds Averaged Navier–Stokes (RANS) methods, where the ensemble averaging tends to remove the unsteady part. RANS models generally perform satisfactorily in less complex flows, whereas, in more complex scenarios, it may result in an unsatisfactory performance. A completely different approach is represented by the so-called large eddy simulation (LES) method, where the large-scale energy-containing eddies are solved directly, while the effects of smaller-scale eddies are simply modeled, resulting in more expensive models than RANS models in terms of computational cost, but results in a remarkable increasing regarding the accuracy of the predictions. In this Special Issue, a total of 11 papers have been published, focused on different disciplines of fluid mechanics, showing the latest advances in each field.

Considering the above, this Special Issue was introduced to collect the latest research addressing the present challenges in fluid mechanics. There were 12 papers submitted to this Special Issue, and 11 papers were accepted (i.e., a 91.6% acceptance rate). When looking back to the papers, various topics have been addressed, mainly based on numerical modelling applied to different approaches, such as turbomachinery, airfoils, offshore renewables, drainage, and other completely different applications from heat transfer and PIV measurement techniques to hot gases and plasmas.

There are three papers focused on rotational machines. The first one, authored by Li, X. et al. [1], presents flow control effects over the geometry of the end wall surface of axial compressors that illustrates the possible variation of end wall flow according to the Bernoulli effect. In the circumferential direction, the full-area unit generates an upslope surface, whereas, on the pressure side of the end wall, all side effects are contrary to the suction side. The second paper, authored by Tan, L. et al., provides another review on the effect of blade wrap angle on the hydrodynamic radial force of a single blade centrifugal pump [2] through numerical simulation because the dissipation losses show a decreasing trend as the blade wrap angle increases, suggesting that the available blade wrap angle for the pump should be in a well-defined angle, to achieve a better hydraulic performance and stable flow field. Finally, in the third paper, Bengoechea, A. et al. [3] presented an approach to the flow pattern of axial fans via a compact model based on three directional pressure loss coefficients comparing several flow patterns obtained through different modeling strategies, highlighting a porous media; thus, a significant reduction in the time needed to create the mesh as it is not necessary to generate the geometries of the pores, being a great advantage.

Additionally, there are another three papers focused on different types of measurements and heat transfer techniques. The first one by Kotšmíd, S. et al. [4] introduced a theoretical analysis of heat transfer in a heated tube bank, providing the most suitable variant for a unique reference temperature in terms of a constant value for all tube angles and several Reynolds number ranges, which are in good agreement with the most frequently

Citation: Blanco, J.M. Special Issue on the Advances in Fluid Mechanics. *Appl. Sci.* **2023**, *13*, 5492. https://doi.org/10.3390/app13095492

Received: 4 March 2023
Accepted: 25 April 2023
Published: 28 April 2023

used correlating equations. In the second paper, authored by Yazdi, S.G. et al. [5], particle image velocimetry (PIV) was used to capture the fluid dynamics and velocity reduction within silicone aneurysm replicas. A circulatory mimicking loop was developed to pump the flow through the silicone models. While wall shear stress is an important parameter in the analysis of aneurysm flow diversion effects, it was not calculated in this study due to light reflection at the lumen wall that may lead to error in the cross-correlation algorithm, whereas CFD analysis will be implemented soon in three dimensions and high resolution. Finally, the third paper, authored by Basarab, M. et al. [6], presents a novel approach for solving CFD problems in the thermal accelerometer's cavity, which is based on the combined use of Rvachev's R-functions method and the Galerkin technique. Different bases were applied in this work, both spectral (polynomial) and local (B-splines), and good results were achieved for fields evaluated in domains of simple geometry without localized inhomogeneities.

Two additional papers were focused on special fluids, such as plasmas and hot gases. The first one, authored by Regodón, G. et al. [7], highlighted the validity of the orbital theories that model the ion current collected by a cylindrical Langmuir probe immersed in low-pressure, low-temperature helium plasma. Therefore, a transition from the validity of the orbital theories towards the radial theories is expected, depending on the experimental conditions of the plasma. Tang, W. et al., in their paper "Performance and Modeling of a Two-Stage Light Gas Gun Driven by Gaseous Detonation" [8], addressed a two-stage light gas gun driven by gaseous detonation. This type of model is required because different tests are frequently conducted in experimental facilities, and high-quality simulations are needed. The proposed quasi-one-dimensional model provides accurate simulation results of the internal ballistic process of the light-gas gun.

Song, Y. et al. authored the paper "Development of Driftwood Capture Trellis for Capturing Driftwood in Agricultural Drainage Ditches" [9], dealing with driftwood capture trellis for agricultural drainage ditches, contributing to a reduction in overflow damage caused by driftwood sedimentation evaluated the performance of the system through hydraulic experiments, which will be useful in mitigating overflow damage by capturing driftwood in agricultural drainage ditches before it flows into the drainage.

As an emerging renewable energy technology, there is intensive research on wave/marine energy, but more should come. In this Special Issue, a paper put the focus on this technology, authored by Galera-Calero, L. et al. [10], where a detailed study is undertaken with the computational modelling of a sub-platform for floating offshore wind using the software Star-CCM+ with the application of the RANS approach that allowed to reduce the computational cost of each simulation by 50%. The model developed in this paper can be adopted for further study in wave energy.

Lastly, the paper 'Numerical Simulation of Random Cavitation Suppression Based on Variable NACA Airfoils' [11] introduces a deformable covering in the cavitation-prone area of a particular airfoil, which can be changed adaptively to meet the requirement of suppressing random cavitation. The paper was authored by Shi, W. et al., where authors first highlight the challenges for the commonly used Smagorinsky model of the large Eddy simulation (LES) method and the cavitation model of the viscosity-modified model, which can accurately predict the cavitation shedding frequency at the wake of the hydrofoil (which is the main cause of cavitation shedding).

Different examples of the application of the latest advances of Fluid Mechanics have been addressed in this Special Issue, highlighting the true relevance of such improvements in many relevant fields through a common bond, represented here by the implementation of the computational fluid dynamic techniques capable of simulating very complex situations that otherwise would have been impossible both to visualize and to obtain verifiable results. The future of these techniques is very promising as, in fact, most of the pre-conceptual industrial designs nowadays cannot be conceived without the application of such advanced techniques.

Acknowledgments: This issue would not be possible without the contributions of various talented authors, hardworking, and professional reviewers. Congratulations to all authors—no matter what the final decisions of the submitted manuscripts were—the feedback, comments, and suggestions from the reviewers and editors helped the authors to improve their papers. We would like to take this opportunity to record our sincere gratefulness to the research group IT1514-22. Finally, we place on record our gratitude to the editorial team of *Applied Sciences*.

Conflicts of Interest: The authors declare no conflict of interest.

References

1. Li, X.; You, F.; Lu, Q.; Zhang, H.; Chu, W. The Investigation of a New End Wall Contouring Method for Axial Compressors. *Appl. Sci.* **2022**, *12*, 4828. [CrossRef]
2. Tan, L.; Yang, Y.; Shi, W.; Chen, C.; Xie, Z. Influence of Blade Wrap Angle on the Hydrodynamic Radial Force of Single Blade Centrifugal Pump. *Appl. Sci.* **2021**, *11*, 9052. [CrossRef]
3. Bengoechea, A.; Antón, R.; Rivas, A.; Larraona, G.S.; Ramos, J.C. Compact Model of a Screen under Fan-Induced Swirl Conditions Using a Porous Media Approach. *Appl. Sci.* **2021**, *11*, 1999. [CrossRef]
4. Kotšmíd, S.; Brodnianská, Z. Determination of the Reference Temperature for a Convective Heat Transfer Coefficient in a Heated Tube Bank. *Appl. Sci.* **2021**, *11*, 10564. [CrossRef]
5. Yazdi, S.G.; Mercier, D.; Bernard, R.; Tynan, A.; Ricci, D.R. Particle Image Velocimetry Measurements of the Flow-Diverting Effects of a New Generation of the eCLIPs Implant for the Treatment of Intracranial Bifurcation Aneurysms. *Appl. Sci.* **2020**, *10*, 8639. [CrossRef]
6. Basarab, M.; Giani, A.; Combette, P. Thermal Accelerometer Simulation by the R-Functions Method. *Appl. Sci.* **2020**, *10*, 8373. [CrossRef]
7. Regodón, G.F.; Díaz-Cabrera, J.M.; Fernández Palop, J.I.; Ballesteros, J. Influence of the Ion Mass in the Radial to Orbital Transition in Weakly Collisional Low-Pressure Plasmas Using Cylindrical Langmuir Probes. *Appl. Sci.* **2020**, *10*, 5727. [CrossRef]
8. Tang, W.; Wang, Q.; Wei, B.; Li, J.; Li, J.; Shang, J.; Zhang, K.; Zhao, W. Performance and Modeling of a Two-Stage Light Gas Gun Driven by Gaseous Detonation. *Appl. Sci.* **2020**, *10*, 4383. [CrossRef]
9. Song, Y.; Park, M. Development of Driftwood Capture Trellis for Capturing Driftwood in Agricultural Drainage Ditches. *Appl. Sci.* **2020**, *10*, 5805. [CrossRef]
10. Galera-Calero, L.; Blanco, J.M.; Iglesias, G. Numerical Modelling of a Floating Wind Turbine Semi-Submersible Platform. *Appl. Sci.* **2021**, *11*, 11270. [CrossRef]
11. Shi, W.; Shi, Z.; Xie, Z.; Zhang, Q.; Yang, Y.; Tan, L. Numerical Simulation of Random Cavitation Suppression Based on Variable NACA Airfoils. *Appl. Sci.* **2021**, *11*, 11618. [CrossRef]

applied sciences

Article

The Investigation of a New End Wall Contouring Method for Axial Compressors

Xiangjun Li [1], Fuhao You [1,*], Qing Lu [1], Haoguang Zhang [2] and Wuli Chu [2]

1 Naval Architecture and Ocean Engineering College, Dalian Maritime University, Dalian 116026, China; xjli@dlmu.edu.cn (X.L.); lu18438606272@163.com (Q.L.)
2 School of Power and Energy, Northwestern Polytechnical University, Xi'an 710072, China; zhg@nwpu.edu.cn (H.Z.); wlchu@nwpu.edu.cn (W.C.)
* Correspondence: fhyou@dlmu.edu.cn

Abstract: To further control corner separation in high-load axial compressors, this study proposes a new end wall contouring method. It defines multiple standard "surface units" with particular flow control effects and then applies a linear combination, finally forming the geometry of the end wall surface. Based on design experiences, three different end wall contouring cases are generated and calculated on a high-load compressor cascade in the first step. The results show that the new method achieves a clear and intuitive influence on the end wall geometry, with a proper number of design variables, and can effectively combine variables with the development of secondary flow. In the second step, the new method was applied to an axial compressor, with an improvement in the design variables. Although the end wall contouring only improved the efficiency of the compressor stage on the right part of its operating map, the experimental results of the flow field show that the corner separation and end wall loss are suppressed at multiple inflow conditions. The results thus verified the practical effect of the newly developed end wall contouring method.

Keywords: corner separation; end wall contouring; flow control; compressor

Citation: Li, X.; You, F.; Lu, Q.; Zhang, H.; Chu, W. The Investigation of a New End Wall Contouring Method for Axial Compressors. *Appl. Sci.* **2022**, *12*, 4828. https://doi.org/10.3390/app12104828

Academic Editor: Jesús María Blanco

Received: 14 February 2022
Accepted: 5 May 2022
Published: 10 May 2022

1. Introduction

Compressors of the turbine engine have been designed with higher loads in recent years. As a result, the high pressure gradient across the blade channel increases the three-dimensional effect of the end wall region. The resultant corner separation becomes a problem for the efficiency and stability of compressors. Therefore, the high-load compressor commonly requires flow control techniques to improve its performance.

As a passive flow control technique in turbomachinery, end wall contouring was first proposed in turbines and then applied in axial compressors in past decades [1]. The basic principle of end wall contouring is to alter the local static pressure field by generating small concave and convex on the end wall surface. The variation of the pressure field then affects the secondary flow motion, finally helping to relieve the corner separation. Compared to the well-known three-dimensional sweep and lean, end wall contouring has little effect on the stacking law of the blade and barely alters the blade loading. In recent years, numerous numerical and experimental studies have confirmed its significant effect in suppressing corner separation and improving the performance of compressors [2–6].

The end wall contouring design method broadly comprises two categories in published studies. One is the empirical method. The design process uses the analytical formula or manual adjustment to create the geometry of the entire end wall, thereby controlling the overall end wall secondary flow. According to the Bernoulli equation, a convex surface forms a convergent channel locally and depressurize the flow, and a concave surface has the opposite effect. Thus, a downslope from the PS to the SS (called DPS for short) will reduce the cross-passage pressure gradient and suppress the end wall secondary flow. Hu's research [7] generated the DPS with a sine function to suppress the secondary flow, finally reducing the corner separation and increasing the peak efficiency by 0.45%. The DPS is also found in a much more recent study of a compressor cascade by Cao [8], where the end wall contouring similarly reduced 18% of the total pressure loss. An afterward study by Meng et al. [2] combined the end wall contouring with the SS corner profiling, proving that even using a straight-shaped DPS would suppress the end wall second flow and benefit the separation control. Note that the DPS is not the only contouring rule for loss control. The exception is in Harvey's research [9], where the end wall contouring is manually constructed with an upslope from PS to the SS (called UPS for short), thus accelerating the end wall secondary flow to pre-mix with the accumulated low-energy fluid. Finally, the spanwise distribution of the loss coefficient was reduced by up to 7% relative to the original case. Therefore, it is hard to say whether decelerating or accelerating the end wall secondary flow would be more effective in separation control.

The other end wall contouring method, i.e., the optimization method, shows more diversity than the empirical method. The parametric end wall surface is traditionally constructed using a lofted surface over dozens of uniformly distributed free control points, thus enabling highly flexible variation to work out the best end wall contouring in the design space. According to the optimization results, some research found weakening of the end wall secondary flow to be most effective in controlling corner separation [10–15]. However, the results are different from the DPS in the empirical method. The end wall might have a downslope, but it is more likely a localized surface near the SS [10] or from the center of the blade channel [11] rather than a full-range structure. Some even find that the optimum end wall generates a streamwise groove, thus inducing a contour vortex to the passage vortex and deflecting the end wall secondary flow before it reaches the SS corner [12–15]. The most frequently reported effective flow control is the acceleration of cross and climbing flow in the SS corner [3,16–19]. The optimization results show their diversity even with a common feature. In Harvey's study [16], the acceleration of the end wall secondary flow is caused by a localized fillet-like upslope surface to the SS corner. Varpe's research [17] shows a similar feature but a different position and area. Moreover, in the studies of Zhang [11], Reising [18], Lepot [19], and Ma [3], the upslope surfaces in the SS corner are all connected with concave surfaces in the peripheral regions and thus able to enhance the end wall secondary flow further.

Table 1 compares the general effect of the flow control technique on the compressors or turbines in some typical studies. There are two studies of end wall contouring included, showing that the improvement of efficiency reaches 0.3~0.5%, which is comparable to those of the three-dimensional blading. It is worth noting that the above summary only describes the general features and effects of the end wall surface. More details, including the location, intensity, and area of the effect of end wall contouring, are not mentioned, but they may change from one case to another. According to some more recent research, the diversity of the optimum results should be associated with the variation of incidence of inflow [4], the solidity of the blade passage [5], or even the thickness and skewness of the inflow boundary layer [6,20]. Therefore, the results actually indicate that the parametric surface of end wall contouring should, at the very least, affect the end wall secondary flow from more than one position to make itself more effective. This is what the traditional empirical method lacks. In comparison, the optimization method seems more reliable, but there are still problems during application. Except for its time-consuming simulation during the design iterations, the flow control rules of a particular optimal end wall contouring are commonly not widely applicable. Thus, the design rules of the end wall contouring are difficult to establish. With the above problems, this study developed a new end wall contouring method for compressors, which shows improvement in the following aspects:

1. The number of parameters is limited to an appropriate level, making the method easy to use. The parameters have a clear and intuitive influence on the end wall geometry and the intent of flow control.
2. The design space is large enough to accommodate suitable aerodynamic end wall shapes for a wide range of compressor cases.

3. The new method can take into account the control of multiple local secondary flows while facilitating the integration of previous design experience.

Table 1. Effect of different flow control techniques on turbomachineries.

Flow Control Techniques	Application	Researchers	Improvement of Efficiency
fillet in the SS corner	compressor rotor 67	Ji [21]	0.3%~0.5% (different rotating speed)
three-dimensional blading	a compressor rotor	Mahmood [22]	0.7% (peak efficiency)
three-dimensional blading	compressor stage 35	Cheng [23]	0.53% (peak efficiency)
end wall contouring	a compressor stage	Sun [24]	0.2%~0.3% (best improvement)
end wall contouring	Trent 500 HP turbine	Brennan [25]	0.4% (peak efficiency)
end wall contouring	a compressor rotor	Hu [7]	0.45% (peak efficiency)

The following paragraph will introduce the new end wall contouring method in Section 2, followed by a numerical investigation of a linear cascade in Section 3. An experiment on a compressor stator was carried out to verify the new method, and the results are presented in Section 4. Finally, Section 5 concludes the paper.

2. New End Wall Contouring Method

To control the second flow in multiple local areas, the main idea of the new end wall contouring method is to generate multiple "units" with particular effects for the secondary flow. Then, we apply a weighted superposition to these units, as shown in Figure 1, to obtain a comprehensive flow control effect. The parametric equation for the end wall surface can be written as

$$R(z,\theta) = \sum_{i=1}^{n} C_i f_i(z,\theta) \tag{1}$$

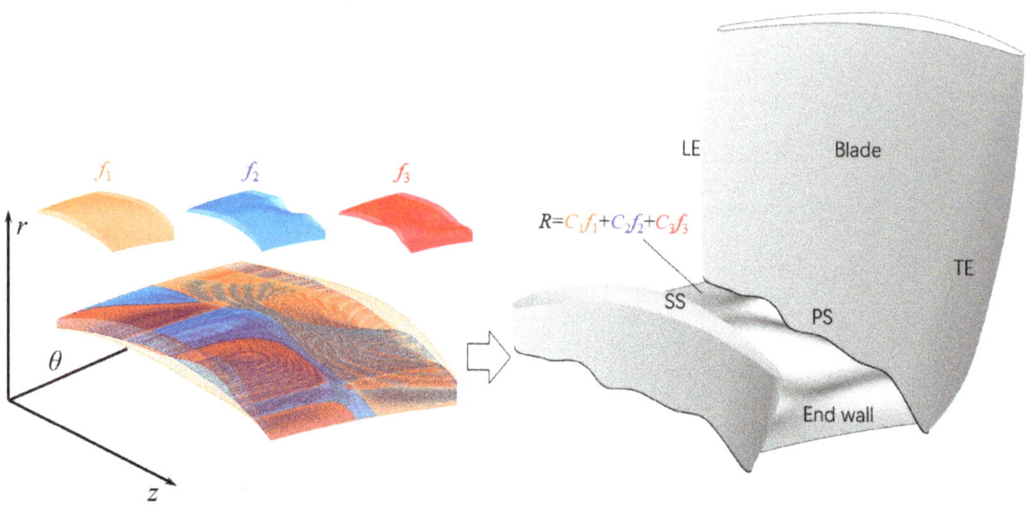

Figure 1. Basic idea of the new end wall contouring method.

Each $f_i(z, \theta)$ term of Equation (1) is a smooth and continuous function defined on the end wall z-θ plane, meaning the spanwise coordinate (r) of an individual end wall contouring unit. Herein, they are named unit functions for brevity in the following statement. We defined C_i as the corresponding weight factor. The overall shape of end wall contouring will be dependent on each unit function and the value of their weight factors. Considering the previous research in the Introduction, we propose two types of end wall contouring units.

2.1. The Definition of the End Wall Contouring Units

(1) The full-area unit

The first type of end wall contouring unit is designed to accelerate or decelerate the cross-passage secondary flow over the end wall region. It is termed "full-area unit" for brevity. The definition is sketched in Figure 2. For the convenience of presentation, Figure 2a maps the z-θ end wall surface of the blade channel to a dimensionless 1×1 standard space and sets η, ξ, and ε to represent the spanwise, circumferential, and streamwise directions of the compressor channel. $\varepsilon = 0$ and $\varepsilon = 1$ represent the leading edge (LE) and trailing edge (TE) lines; $\xi = 0$ and $\xi = 1$ represent the PS and SS. Then, the control equations of the full-area unit can be expressed as

$$f(\varepsilon, \xi) = A(\varepsilon)F(\xi) = -\cos(\pi(\varepsilon - 0.5))^3 \sin(\pi(\xi - 0.5)) \tag{2}$$

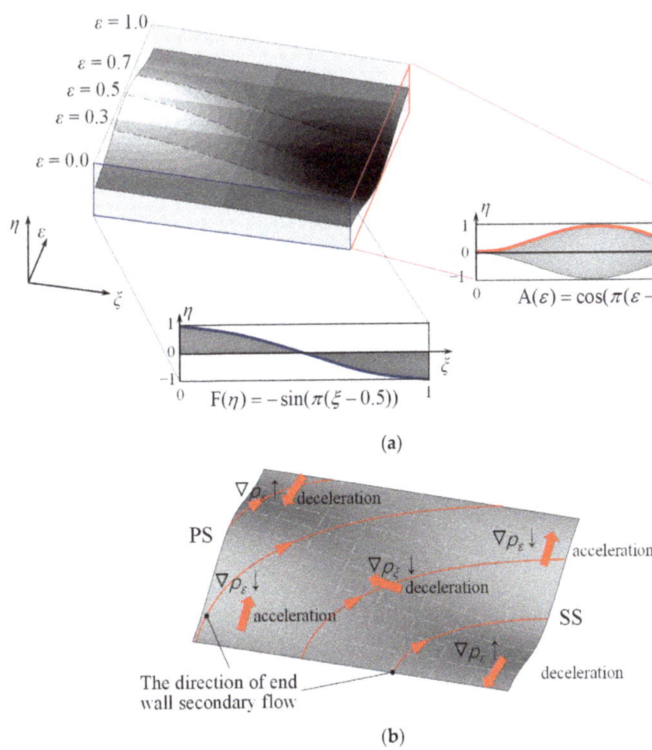

Figure 2. The full-area unit. (**a**) Definition of the full-area unit; (**b**) The effect on the pressure field and the end wall secondary flow.

Note the streamwise function, i.e., $A(\varepsilon) = \cos(\pi(\varepsilon - 0.5))^3$ acts as the amplitude of the transverse function $F(\xi) = -\sin(\pi(\xi - 0.5))$, and the maximum value of $A(\varepsilon)$ locates at

$\varepsilon = 0.5$. To further smooth the surface, we selected five axial sections (transparent sections labeled from S1 to S5 in Figure 2) at the streamwise position of $\varepsilon = 0, 0.3, 0.5, 0.7$, and 1 in the full-area unit to construct the lofting surface. Considering that there are only five sections, the interpolation of the surface uses the cubic spline to ensure it accurately gets through all the sections, finally forming the geometry for the full-area unit.

Consider that when the full-area unit is mapped to the actual end wall of the axial flow compressor, the shape of the surface causes the pressure side of the blade channel to sink and the suction side to rise, with the maximum slope at the 0.5 axial chord length (c_a) position. Figure 2b thus illustrates the possible variation of end wall flow according to the Bernoulli effect. In the circumferential direction, the full-area unit generates an upslope surface from the SS to the PS, just as in previous research [7,8]. So, the transverse pressure gradient (∇p_ξ) will decrease, and the cross-passage secondary flow will decelerate in the entire area. There are some side effects. On the suction side of the end wall, the variation of the streamwise pressure gradient (∇p_ε) will stay positive from the LE to the mid-chord and then negative till the TE. Therefore, the streamwise component of end wall flow will decelerate in the front half and then accelerate in the rear half. On the pressure side of the end wall, all side effects are contrary to the suction side. It should also be noted that when applied to an actual compressor, all the above flow control, including its impact on the cross-passage secondary flow and the side effects, may turn in the opposite direction when the weight factor is less than zero.

(2) The localized unit

The second type of end wall contouring unit is defined to facilitate the secondary flow control in one or more particular local areas in the blade channel. It is termed the "localized unit" for brevity. The definition of this end wall contouring unit is shown in Figure 3a. Different from the full-area unit, the streamwise starting and ending positions ε_1, ε_2, and circumferential starting and ending positions ξ_1, ξ_2 in the localized unit are all free parameters. Varying these parameters will change the shape of the localized unit.

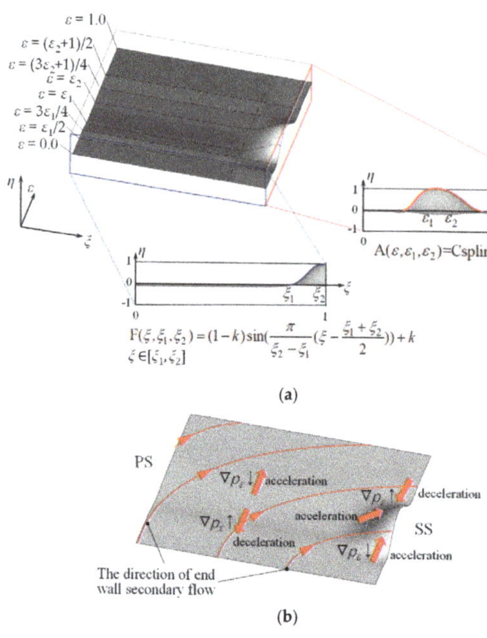

Figure 3. The localized unit. (**a**) Definition of the localized unit; (**b**) the effect on the pressure field and the end wall secondary flow.

The parametric equation of the localized unit is

$$f(\varepsilon,\xi) = (F(\xi,\xi_1,\xi_2) + k)A(\varepsilon,\varepsilon_1,\varepsilon_2) \quad (3)$$

where

$$A(\varepsilon,\varepsilon_1,\varepsilon_2) = \text{Cspline}(E_i, H_i) \quad (4)$$

and

$$F(\xi,\xi_1,\xi_2) = \begin{cases} 2k-1, \ \xi < \xi_1 \\ (1-k)\sin\left(\frac{\pi}{\xi_2-\xi_1}\left(\xi - \frac{\xi_1+\xi_2}{2}\right)\right) + k, \ \xi_1 < \xi < \xi_2 \\ 1, \ \xi > \xi_2 \end{cases} \quad (5)$$

The function of $A(\varepsilon, \varepsilon_1, \varepsilon_2)$ of Equation (4) represents a cubic spline interpolation over eight control points and defines the amplitude distribution in the streamwise direction. The eight points are $\{E_i, H_i\} = \{(0, 0), (\varepsilon_1/2, 0), (3\varepsilon_1/4, 0), (\varepsilon_1, 1), (\varepsilon_2, 1), ((3\varepsilon_2 + 1)/4, 0), ((\varepsilon_2 + 1)/2, 0), (1, 0)\}$. The factor k is the zeroing factor. It is defined as

$$k = k \left| \int_{\xi_1}^{\xi_2} F(\xi,\xi_1,\xi_2) = 0 \right. \quad (6)$$

Thus, k is able to make $F(\xi)$ integrate to 0 on the transverse region of $\xi \in [0, 1]$, thus ensuring that the cross-sectional area of the blade channel does not change.

Figure 3b shows the possible variation of the pressure field and the end wall flow when applying a localized unit in the end wall contouring. If $C_i > 0$, the end wall surface would warp up and create an upslope from ξ_1 to ξ_2. Thus, the transverse pressure gradient within $\{\xi \in [\xi_1, \xi_2] \cap \varepsilon \in [\varepsilon_1, \varepsilon_2]\}$ will increase, thus accelerating the cross-passage secondary flow locally. The side effects are similar to the full-area unit. On the suction side of the profiling region, the variation of ∇p_ε will accelerate the streamwise end wall flow in the upstream region and decelerate it in the downstream area. On the pressure side, the side effects become the opposite. As in the full-area unit, all the flow control may turn to the opposite direction when $C_i < 0$.

2.2. Generating End Wall Contouring in the Standard Space

The above end wall contouring units are combined according to Equation (1) to obtain geometry within the standard space. Note that each of the above units would affect end wall secondary flow in a particular region. Thus, the physical meaning of combining them as Equation (1) is to superimpose their effects, finally obtaining an end wall contouring with a comprehensive control effect on the end wall secondary flow. Supposing an end wall contouring includes a total of n units (labeled as f_1 to f_n), i.e.,

$$H = C_1 f_1 + C_2 f_2 + C_3 f_3 \ldots + C_n f_n \quad (7)$$

The weight factor C_i would represent how much the ith unit affects the final control force on the end wall secondary flow.

Figure 4 illustrates an end wall surface consisting of a full-area unit and a localized unit as an example. The former is designed to suppress the cross trend of secondary flow over the end wall range, while the latter is intended to promote the cross trend of secondary flow in the SS corner region. As can be seen from the figure: the changes induced by adjusting the C_1 and C_2 are significant. When $C_2/(C_1 + C_2)$ changes from 0.1 to 0.9, the overall slope of the surface from SS to PS gradually weakens, and the local shape along the SS corner becomes more and more prominent. Although the linear combination of the above two end wall contouring units does not mean their flow control would have a linear combination, it is known from the final end wall surface that the effect of the localized unit on the end wall flow near the SS corner gradually increases with the value of C_2/C_1.

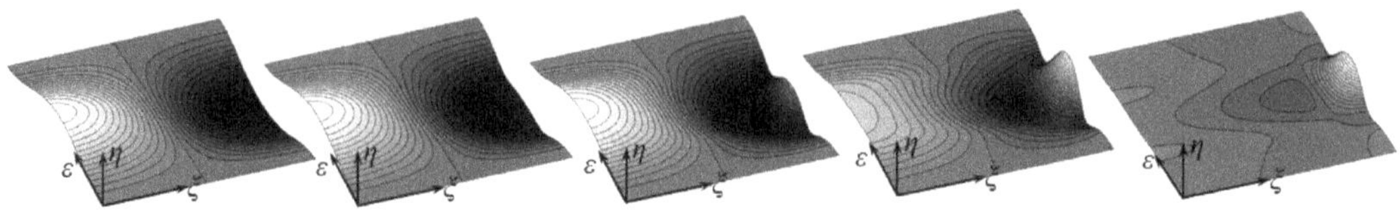

Figure 4. The effect of weight factor on the end wall contouring.

2.3. Generating End Wall Contouring for the Actual Compressor

After obtaining the end wall contouring in the standard space of $\{(\varepsilon, \xi, \eta) \mid \varepsilon \in [0, 1], \xi \in [0, 1], \eta \in [-1, 1]\}$, the shape of the end wall contouring can be mapped directly to the three-dimensional end wall region of the axial compressor according to their spatial correspondence.

As shown in Figure 5, the coordinate system of the compressor end wall region is (r, θ, z). r_0 represents the radius of the baseline end wall surface; z_{LE} and z_{TE} represent the LE and TE axial coordinates; θ_{PS} and θ_{SS} represent the circumferential coordinates of the PS and the SS. Here, we correspond the end wall's LE and TE to $\varepsilon = 0$ and $\varepsilon = 1$ and the SS and PS that form the blade channel to $\xi = 0$ and $\xi = 1$, and we limit the waviness of the actual end wall contouring within $\pm\Delta R$. Then, the end wall contouring in the standard space can be mapped to the actual axial compressor through

$$\begin{cases} r = \eta \Delta R(z) + r_0(z) \\ z = \varepsilon(z_{TE} - z_{LE}) + z_{LE} \\ \theta = \xi(\theta_{SS}(z) - \theta_{PS}(z)) + \theta_{PS}(z) \end{cases} \tag{8}$$

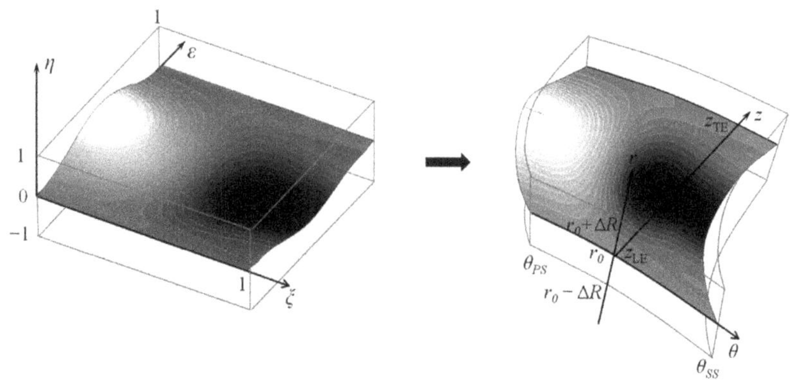

Figure 5. Mapping from the standard space to the end wall region of a compressor passage.

The above section discusses the definition and implementation process of this new end wall contouring method. Note that the new method can theoretically create arbitrary shapes by combining an infinite number of end wall contouring units. However, some preliminary studies indicate that combining one full-area unit and two localized units will usually be sufficient to control the corner separation. So, the new method generally requires less than 12 parameters for all the design processes.

3. Application in a High-Load Compressor Cascade

To evaluate the performance of the new end wall contouring method, it was first applied on a compressor cascade to see how it controls the corner separation.

3.1. The Baseline Cascade and CFD Method

The research object is a high-load linear compressor cascade with a designed flow turning angle of 52°. The aspect ratio of the cascade is 2.0, and the solidity exceeds 2.1. There are detailed experimental studies that have been carried out on this cascade [26,27], showing that severe corner separation arises in this cascade even under the design incidence (i = −1°), and it will develop to corner stall when i exceeds 7°. Thus, the working point with i = +7° is named the near-stall (NS) point for brevity.

The computational domain of the cascade only includes a half-span region of the cascade. Figure 6 provides an overview of the computational mesh. The inlet and outlet passages both use long extended "H" blocks. The blade-to-blade flow surface employs refined nodes in the center with the "O4H" topology to achieve higher orthogonality. The number of nodes in some critical sections are also presented. The computational domain extended to 1.2 times the axial chord (C_a) upstream of the LE, where the inlet velocity is taken during the experiment, and to 2.0 C_a downstream of the TE for the uniformity of the outflow parameter. The mesh in the boundary layer region upon the blade surface and the end wall is further refined for accuracy in calculating the corner separation. The growth rate of mesh from the solid wall to the flow field is less than 1.1, and the y^+ is less than two at the first level upon the solid wall. The boundary condition is given according to the experimental results. The inlet boundary defines the static temperature and the inflow velocity. The thickness of the inlet boundary layer is about 12.5%*span*. The outlet boundary defines a uniform static pressure.

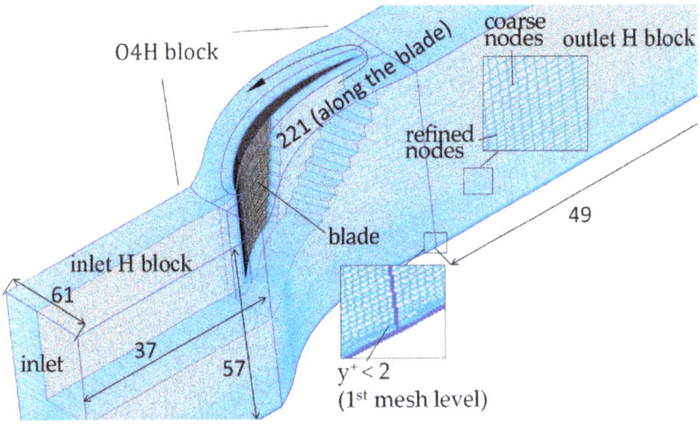

Figure 6. Overview of the computational mesh.

The flow field of the cascade is regarded as compressible and full-turbulent. The *Re* for the current cascade is 1.4 × 10^5. Table 2 provides details of the inflow condition. It is calculated with Numeca Fine/Turbo by solving the compressible steady Reynolds-averaged Navier–Stokes (RANS) equations with the k-ω turbulence model. The spatial discretization scheme uses the second-order central scheme. The convergence criterion was set to a value of 1 × 10^{-6} for the RMS residual values. Local time stepping techniques were applied to speed-up convergence.

Table 2. CFD settings for the inflow condition of the cascade.

Inlet Conditions	Values
static temperature	288.15 K
velocity of mainflow	26.5 m/s
thickness of inlet boundary layer	12.5%*span*
turbulent intensity	0.3%

All the above CFD settings are the same as previous numerical studies on this cascade [4], in which the CFD results were validated using the experimental outflow parameter and the oil-flow test on the SS and end wall surface. A grid convergence study was also applied, showing that a total of 0.75 million grid nodes is sufficient. Here, we offer only the validation of outflow loss coefficient (*Loss*) and flow angle (β) under the design working condition ($i = -1°$), as in Figure 7. The horizontal axis shows the loss coefficient, defined as

$$Loss = \frac{P^* - P^*_{\mathrm{in}}}{P^*_{\mathrm{in}} - P_{\mathrm{in}}} \tag{9}$$

where P^*_{in} and P_{in} represent the inlet total pressure and static pressure in the mid-span position. The vertical axis means spanwise position normalized using the height of the blade. The result labeled "TB" means the working condition with a thicker inlet boundary layer and intensified corner separation. The comparison between the CFD and experimental results shows that the numerical method achieves good accuracy in calculating secondary flow and loss increase when intensified corner separation occurs.

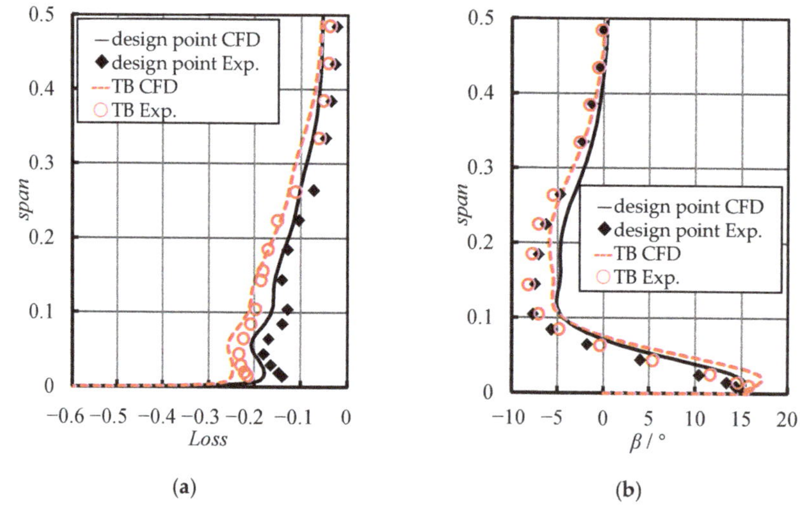

Figure 7. The experimental and CFD results of the pitch-averaged performance at 40% C_a downstream of the TE at $i = -1°$. (**a**) Total pressure loss coefficient; (**b**) Outflow angle.

3.2. Results and Discussion

To apply the new method to the cascade, a software named *PEWs* was developed based on the newly proposed method and was used in this study. This software has a graphical interface for the design variables and enables up to 10 independent units for the end wall contouring. Three different groups of end wall contouring were designed using *PEWs*. Table 3 shows their main parameters. Figure 8 provides an overview of their geometry. The contours show the ratio between the end wall contouring height and the blade height ($\overline{H}$).

Table 3. The design parameter of the end wall contouring cases.

Cases	Parameters of the Full-Area Units		Parameters of the Localized Units					
	Number	w	Number	ε_1	ε_2	ξ_1	ξ_2	w
c1	1	1	0	\	\	\	\	\
c2	1	0.5	1	0.4	0.9	0.9	0.1	0.5
c3	1	0.33	1	0.4	0.9	0.9	0.1	0.67

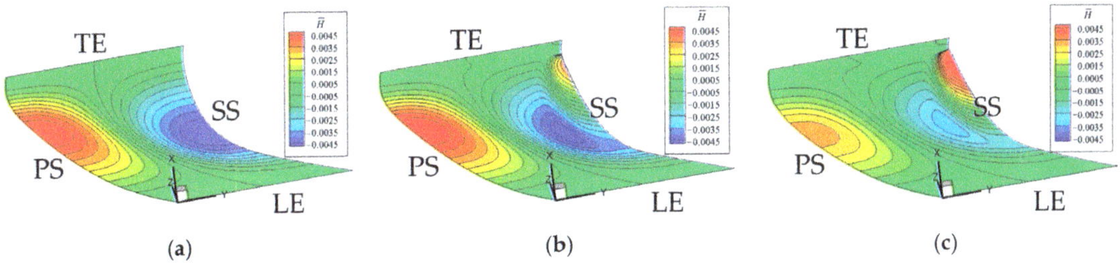

Figure 8. The geometry of the end wall contouring cases: (**a**) case c1; (**b**) case c2; (**c**) case c3.

The first case simply uses a full-area unit. The geometry of end wall contouring forms an upslope from the SS to the PS. The aim is to reduce the transverse pressure gradient of the end wall region and decelerate the cross-passage secondary flow over the end wall range. This case is named c1. The geometry of case c1 is similar to the previously mentioned DPS in Ref. [7]. Thus, it represents the end wall contouring generated using a conventional empirical method.

The second and third cases both adopt a full-area unit and a localized unit to form the end wall contouring, named c2 and c3, respectively, to show their difference. Their design ideas come from previous research [4], which suggests that the end wall contouring should suppress the secondary flow over the end wall surface while accelerating the secondary flow near the SS corner. The difference between the two cases lies in the weight factors. The localized unit of case c3 has a greater weight than case c2, meaning that the latter tends to enhance the secondary flow locally in the SS corner more than the former.

Figure 9 gives the variation of the overall total pressure loss (c_p) with incidence (i) for the three end wall contouring cases. The working condition with $i = -1°$ is the design point. Compared to case c1, case c2 and case c3 improve the performance of the cascade at the design point. Case c2 improves very little, and case c3 reduces the overall loss by about 0.13%; case c1, on the contrary, slightly increases the loss at the design point. As the incidence increases, the benefit of end wall contouring is gradually lost, but the ranking of the three cases is almost unchanged. When i exceeds +3°, all three end wall contouring cases, including case c3, result in a higher loss than the baseline cascade.

Corresponding to the overall performance, Figure 10 provides the spanwise distribution of total pressure loss and flow angle at $0.4C_a$ downstream of the TE. At the design point (Figure 10a), all three cases reduce *Loss* at 0.05*span*, indicating that the three cases have a comparable effect in suppressing the accumulation of low energy fluid at 0.05*span*. The difference is mainly between 0.08*span* and 0.15*span*, where case c1 and case c2 show larger *Loss* than the baseline case, and *Loss* in case c3 is roughly the same as the baseline case. The plots of β show stronger overturning for the baseline case than the end wall contouring cases, indicating that all three end wall contouring cases weaken the end wall secondary flow. Note the phenomenon here: case c3, which reduces the spanwise *Loss* most, does not correspondingly minimize the overturning, while case c1, which increases the spanwise *Loss*, shows the best control of the overturning in the end wall region.

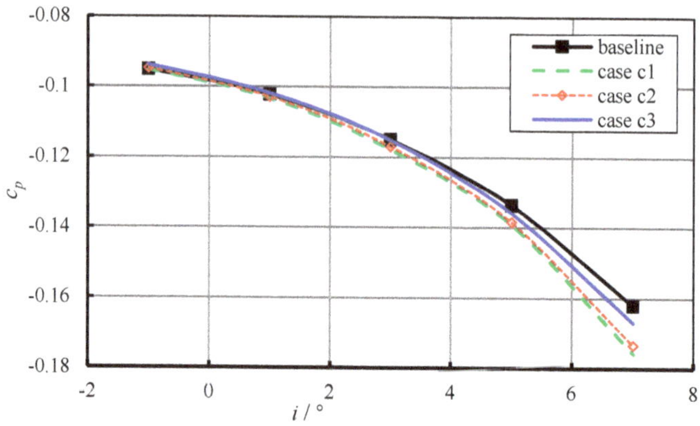

Figure 9. The overall total pressure loss (c_p) for the three end wall contouring cases.

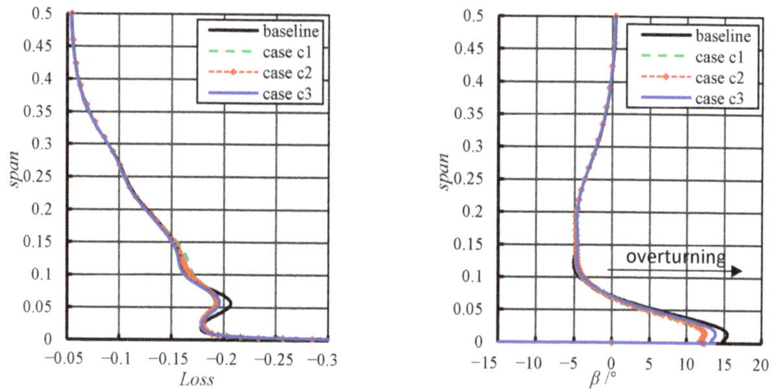

(**a**)

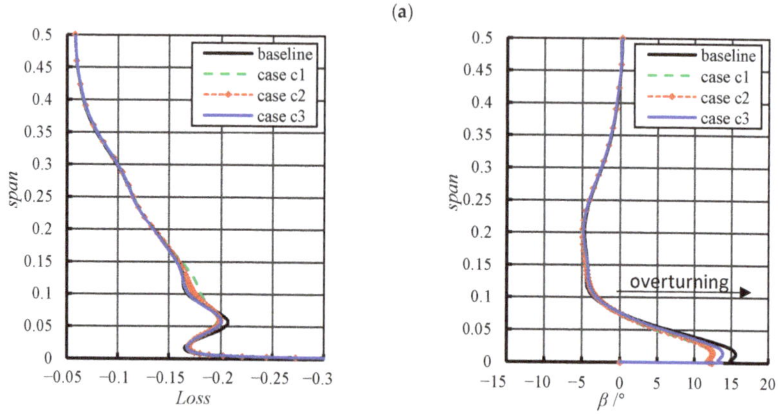

(**b**)

Figure 10. *Cont.*

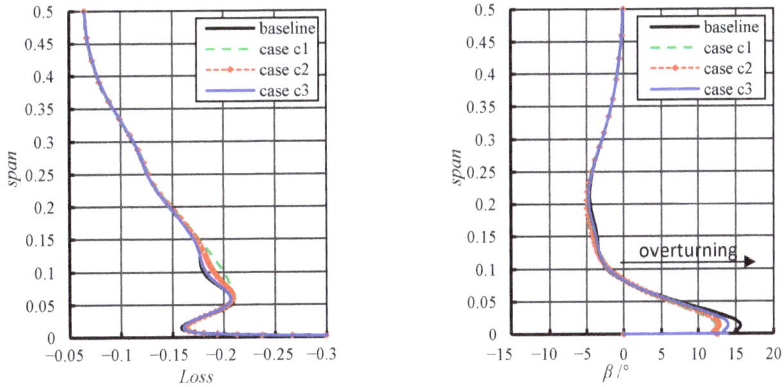

(**c**)

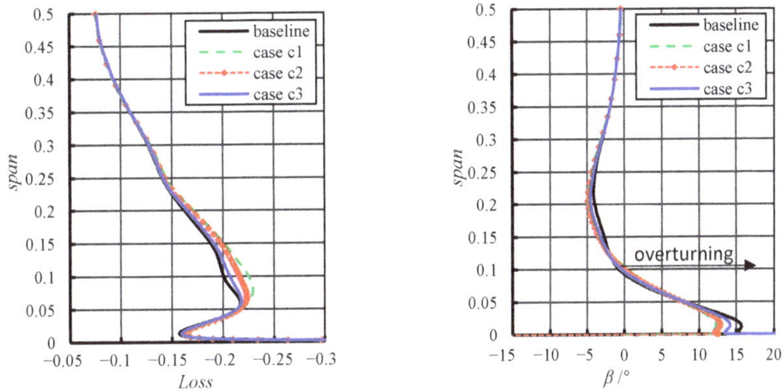

(**d**)

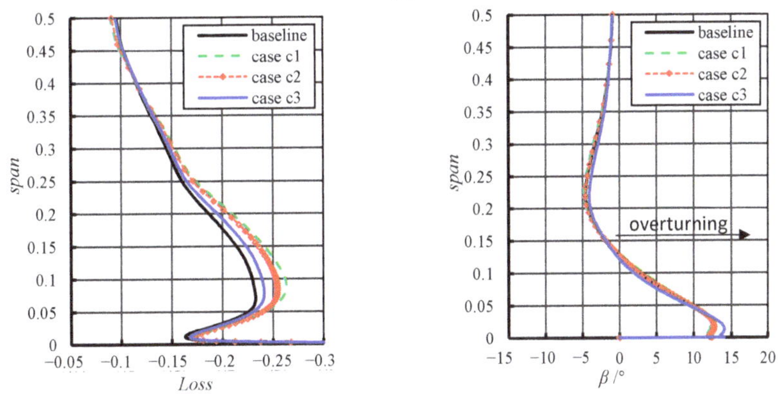

(**e**)

Figure 10. The spanwise distribution of total pressure loss and flow angle at $0.4C_a$ downstream of the TE. (**a**) $i = -1°$, i.e., the design point; (**b**) $i = 1°$; (**c**) $i = 3°$; (**d**) $i = 5°$; (**e**) $i = 7°$, i.e., the near-stall point.

As incidence gradually increases, the benefits of end wall contouring at 0.05*span* decrease, but the increment of *Loss* from the 0.08*span* to 0.15*span* of the end wall contouring cases becomes more significant than the baseline cascade. When the incidence is larger than 3°, the end wall contouring no longer significantly affects the distribution of *Loss* at 0.05*span*, and the phenomenon continues till the NS point ($i = 7°$); between 0.08*span* and

0.15*span*, the increase of *Loss* due to the end wall contouring grows significantly with the incidence. During the incidence increase, all end wall contouring cases show no significant increase of β.

The above results show that the influence of end wall contouring on the cascade mainly works in two areas:

Within 0.05*span*, the relief of the low-energy region due to the end wall contouring is the key to reducing the overall loss. The difference between cases c1, c2, and c3 is not significant, and this benefit gradually disappears with the increase of the incidence.

Between 0.08*span* and 0.15*span*, the different value of total pressure loss results in the difference among cases c1, c2, and c3. With the increase of the incidence, *Loss* of the end wall contouring cases gradually grows to be much higher than the baseline case, leading to a significant drop in their overall performance at large incidence.

In particular, Figure 11 shows the design point and the near stall point, giving the three-dimensional flow field of the baseline cascade and the end wall contouring cases near the end wall region.

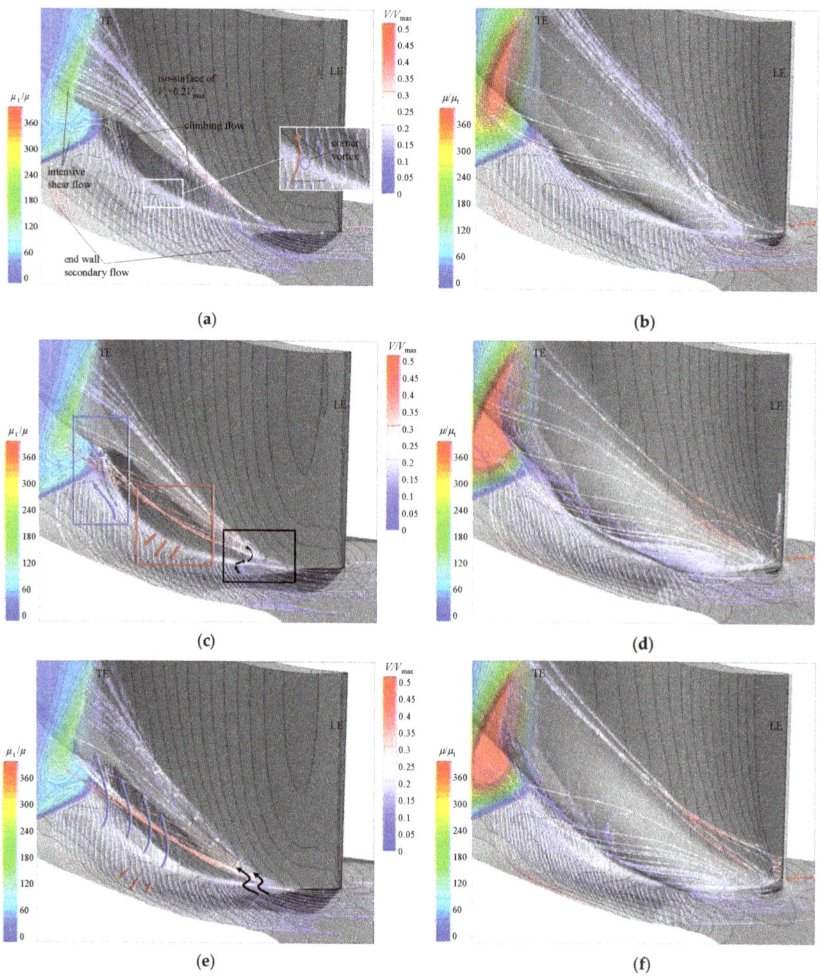

Figure 11. *Cont.*

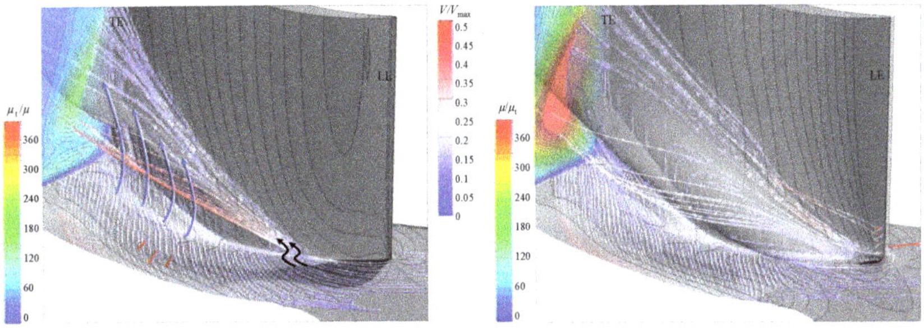

(g) (h)

Figure 11. The three-dimensional flow field of the baseline and end wall contouring cases: (**a**) $i = -1°$, baseline; (**b**) $i = 7°$, baseline; (**c**) $i = -1°$, case c1; (**d**) $i = -1°$, case c1; (**e**) $i = -1°$, case c2; (**f**) $i = -1°$, case c2; (**g**) $i = -1°$, case c3; (**h**) $i = -1°$, case c3.

At the design point (Figure 11a,c,e,g), the transparent grey area in the figures is the iso-surface of $V_z = 0.1V_{\text{max}}$, which indicates the development of separation in the corner region. The contours show the normalized turbulent viscosity (μ_t/μ) within the TE plane, representing the local shear loss caused by separation. The colored curves in the figure are the three-dimensional streamlines in the end wall region, and the color indicates its velocity magnitude. The iso-curve of the static pressure coefficient ($\overline{p}$) is plotted on the end wall and blade surface. $\overline{p}$ is defined as

$$\overline{p} = \frac{p - p_{\text{in}}}{(pt_{\text{in}} - p_{\text{in}})} \tag{10}$$

Figure 11c shows that case c1 significantly weakens the cross-passage secondary flow. The significant changes are mainly reflected in three areas:

1. First, the corner region near $0.2C_a$ to $0.4C_a$ from the LE (in the rightmost black box in Figure 11c), which, according to the streamlines, shows that this is precisely where the corner separation starts. The iso-curves of $\overline{p}$ show that the sinking end wall near the SS corner increases the pressure gradient significantly. Thus, the reverse flow is intensified, and more low-energy fluids accumulate in the corner region. This ultimately increases the local shear loss and exacerbates corner separation.
2. Second, from the mid-chord to the rear region of the SS corner (located in the red box in the middle), the secondary climbing flow weakens, and the same is true for the cross trend of the end wall flow in the outer region (as shown by the red arrows). This is caused by the sinking surface of the end wall on the suction side. The weakened cross trend of end wall secondary flow will inhibit the accumulation of the low-energy fluid and thus help to reduce loss.
3. Third, near the TE of the SS (in the blue box on the left), the end wall iso-curves of $\overline{p}$ show a high-density region, indicating that the streamwise pressure gradient is significantly reduced compared to the baseline case. This should be induced by the local streamwise upslope in the SS corner. This effect enhances the flow momentum of both the end wall flow (shown by the solid blue arrow) and its climbing motion after colliding with the SS (indicated by the dashed blue arrow). The acceleration of the climbing flow mixes with the low-energy fluid on the SS and increases its streamwise momentum. According to the contour of μ_t/μ at the TE plane, this "pre-mixing" effect reduces the shear effect between the separation flow and the main flow and thus brings benefits.

The correlation between the low-energy flow near the TE and upstream streamline is worth noting. According to the streamlines in the baseline cascade, the low-energy flow of the lower span at the TE plane (labeled with a solid white circle in Figure 11a) is associated with the secondary flow in the rear part of the corner region. In contrast, the low-energy flow of the higher span (labeled with a dashed white circle in Figure 11a) comes from the climbing flow along with the SS at the onset of separation in the upstream corner region. Because case c1 exacerbates the corner separation in the upstream SS corner, this indicates that the reduction of the *Loss* within 0.05*span* in Figure 10 comes from the second and third effects mentioned above, while the increase of *Loss* between 0.08*span* and 0.15*span* is mainly due to the first effect.

Compared to case c1, both case c2 and case c3 have less sinking end wall near the SS due to the localized unit. In the SS corner region from $0.2C_a$ to $0.4C_a$, the iso-curves of $\overline{p}$ suggest that the inverse pressure gradient is lower than that of case c1, which makes the local inverse flow induced by case c2 and c3 less severe than that of case c1 (as shown by the black arrows in Figure 11e,g). As a result, the low-energy separation flow at the high span region near TE is also significantly reduced. Another region affected by the localized unit is the corner region from the $0.5C_a$ to the TE. The rise of the end wall near the corner region makes the climbing flow near the TE accelerate significantly and sweep downstream (as shown by the blue arrows in Figure 11e,g). This is consistent with the research in [4]. On the one hand, the energized secondary flow eliminates the low-energy flow in the corner vortex. More importantly, it enhances the "pre-mixing" effect of the secondary flow with the low-energy flow in the corner region mentioned in case c1. Therefore, it is more advantageous than case c1 here, and it can be seen from the TE plane's μ_t/μ contour that the local shear loss is more minor than case c1. The weight factor of the localized unit in case c3 is larger than that of case c2, so the control of pressure gradient in the upstream SS corner and the enhancement of secondary flow pre-mixing near the trailing edge are more significant than that of case c2. The loss reduction in case c3 is, therefore, more significant.

The influence of end wall contouring at the near-stall point ($i = 7°$) is discussed similarly to the design point but more concisely. The flow field is shown in Figure 11b,d,f,h. Compared to the flow field in the design point, the differences between case c1 and the baseline case at the near-stall point are similar to those concluded at the design point. It also appears that the pressure gradient in the SS corner area from $0.2C_a$ to $0.4C_a$ increases significantly, the secondary climbing flow from the SS corner between $0.5C_a$ and TE weakens, and the streamwise pressure gradient near the TE of SS reduces. However, the main difference is that the corner separation at the LE of the cascade is much more severe than the design point (labeled using a dashed white circle in Figure 11d). The local pressure gradient increase due to the sinking end wall intensifies the reverse flow trend, making the corner separation stronger after its generation. This effect amplifies the negative impact of the end wall contouring on the flow field of the cascade between $0.2C_a$ and $0.4C_a$, finally leading to an overall loss increase. Thus the contour of μ_t/μ shows that the low-energy separation flow causes much higher shear losses than the baseline cascade when it develops to the TE.

The advantages of cases c3 and c2 over case c1 are still significant. Under the influence of the localized unit, case c3 and case c2 show more mitigated upstream corner reverse flow and more intense downstream pre-mixing effect of the separation flow than case c1. Therefore, the control of separation loss in cases c2 and c3 is better than case c1. case c3 leads to the slightest increase in loss among all end wall contouring cases due to its larger weight factor of the localized unit than case c2.

In summary, although the corner separation is still not effectively suppressed at the near-stall point, the new end wall contouring method shows its potential for flow control, especially when combined with the previous experience. As already mentioned at the beginning of Section 3, case c1 represents the DPS designed by the traditional empirical method. In comparison, the cases using two end wall contouring units perform better under more than one incidence. What's more, the intended flow control of the full-area and localized units, including the effect of their weight factors, agrees very well with their effects during the application. All these indicate that the new end wall contouring method is superior to the traditional empirical method.

4. Experiment on the Axial Compressor Test Rig

The following will check the performance of the new end wall contouring method in an axial compressor through an experiment.

4.1. The Baseline Compressor Stage and the Design of End Wall Contouring

The compressor of the test rig is a single-stage high-speed axial flow compressor from Northwestern Polytechnic University, as sketched in Figure 12a. The testing stations for the total pressure of the compressor flow field are also illustrated. The diameter of the rotor is 298 mm, with a design speed of 15,200 rpm. Experimental and numerical research on this stage confirmed that when the rotor speeds up to 70% of the design speed, the rotation will reach a relatively steady state for the stator testing, and there will be a considerable corner separation in the casing end wall region of its stator, as shown in the oil-flow photos in Figure 12b. The experiment thus designs end wall contouring on the casing of the stator to improve the performance of the compressor operating at 70% speed.

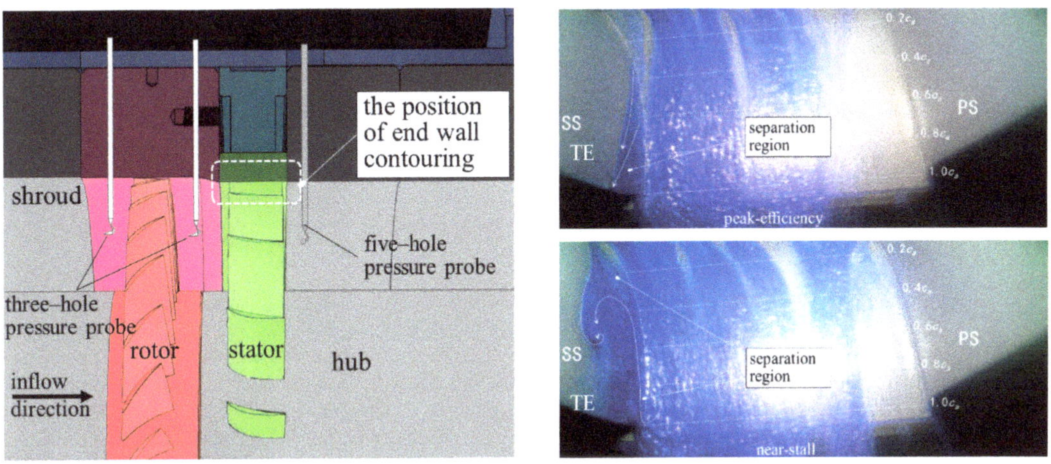

(**a**) (**b**)

Figure 12. The compressor stage for the experiment. (**a**) Test rig; (**b**) Oil-flow in the corner region of casing end wall captured at 70% design speed (top: peak-efficiency point; bottom: near-stall point).

According to the numerical results in the cascade, constructing a concave surface in the front SS corner would raise the pressure gradient and negatively affect the initiation of corner separation, especially when the incidence increases. Therefore, the new design shows an improvement and adopts three units, with the parameters shown in Table 4. The first unit is to suppress the end wall second flow. Here, we use a localized unit with a large profiling area instead of a full-area unit to prevent aggravating the reverse flow in the front SS corner. Its design parameter is listed in the first place of each blank, i.e., $\varepsilon_1 = 0.2$, $\varepsilon_2 = 0.7$, $\xi_1 = 0.5$, $\xi_2 = 1$, and $w_1 = -0.25$. The second localized unit aims to accelerate the secondary climbing flow in the rear part of the SS corner, so the pitch-wise range of the second localized unit is limited to 15% of the circumferential width of the passage (defined with w) beyond the SS in the rear half of the SS. The third localized unit aims to reduce the potential negative effect in the front SS corner area. The range of this unit thus spreads within 30% w from the SS, forming a mild upslope surface to reduce the sinking end wall due to the first localized unit. Figure 13 shows the geometry of the end wall contouring and the photo of the experimental part.

Table 4. The design parameter of the end wall contouring for the stator.

Parameters of the Full-Area Units		Parameters of the Localized Units					
Number	w	Number	ε_1	ε_2	ξ_1	ξ_2	w
0	0	3	0.2; 0.5; 0.1	0.7; 0.8; 0.3	0.5; 0.85; 0.7	1; 1; 1	−0.25; 0.5; 0.25

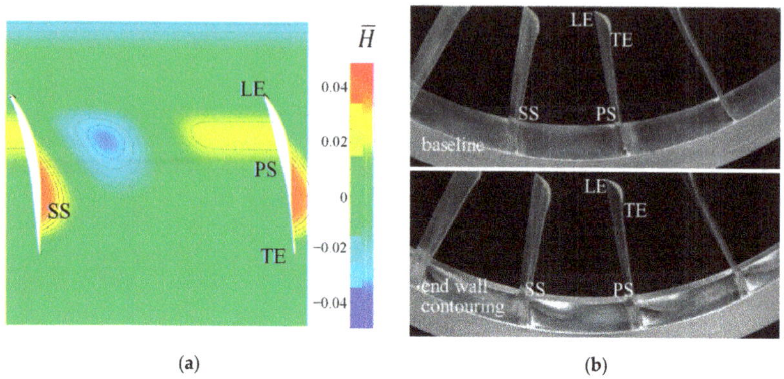

Figure 13. The end wall contouring in the compressor stage (**a**) Normalized end wall contouring height; (**b**) Stator component of baseline and end wall contouring for the experiment.

4.2. Results and Discussion

Figure 14 compares the overall performance of the stage before and after end wall profiling. The experimental results confirmed the positive effect of end wall contouring on the flow field when $\overline{m} > 0.83$. The full-stage efficiency improved by 0.45% at the working point of minimum stator loss (the normalized mass flow rate $\overline{m}$ equals 0.89). Note that the total temperature ratio in Figure 14b is generally the same before and after the end wall contouring, while the variation of the total pressure recovery coefficient (σ) of the stator in Figure 14d is very similar to the efficiency. The variation of σ shows that the total pressure loss in the stator is reduced by 10.44% of its original value. Thus, the change of efficiency should be mainly associated with the stator loss rather than the rotor work. The stall margin does not change much. It is noted that end-wall shaping does not improve the total pressure ratio and efficiency when $\overline{m} < 0.83$. Thus, the specific effect of end wall contouring on the stator casing flow field will be discussed in the next step.

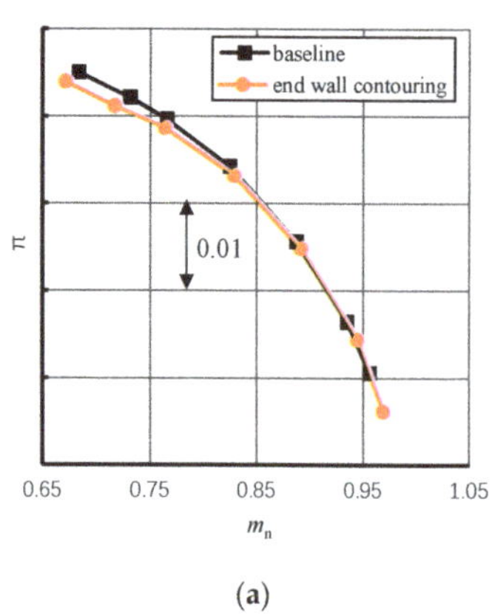

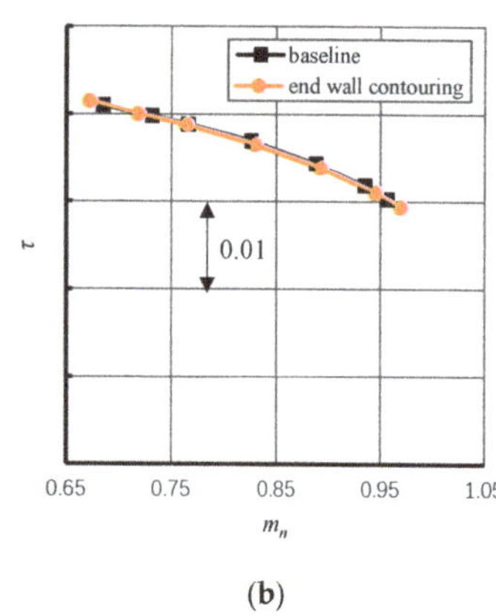

Figure 14. *Cont.*

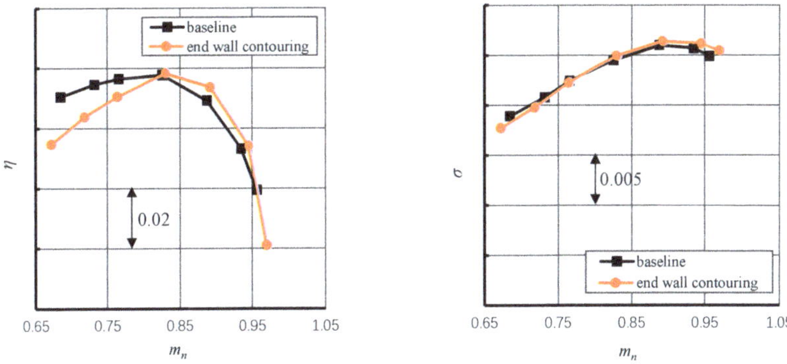

(c) (d)

Figure 14. The overall performance of the compressor stage at 70% design speed. (**a**) Total pressure ratio of the stage; (**b**) Total temperature ratio; (**c**) Efficiency of the stage; (**d**) Total pressure recovery coefficient of the stator (the values of the overall performance are not shown here for confidential reasons. The figure shows only the normalized mass flow rate and the unit of the vertical axis).

Figure 15 selects two specific working points and compares the stators' circumferential averaged total pressure loss before and after end wall contouring. One is the working point with the minimum stator loss ($\overline{m}$ = 0.89). Herein, it is termed the best-matching point for brevity, abbreviated as "BM." Another is a working point during the throttling process ($\overline{m}$ = 0.77), closer to the near-stall point but still relatively stable for the experiment. It is termed the left-map working point in the following passage, abbreviated as "LM." At the LM point, the variation of σ indicates the overall total pressure loss increases by 1.95% of its original value.

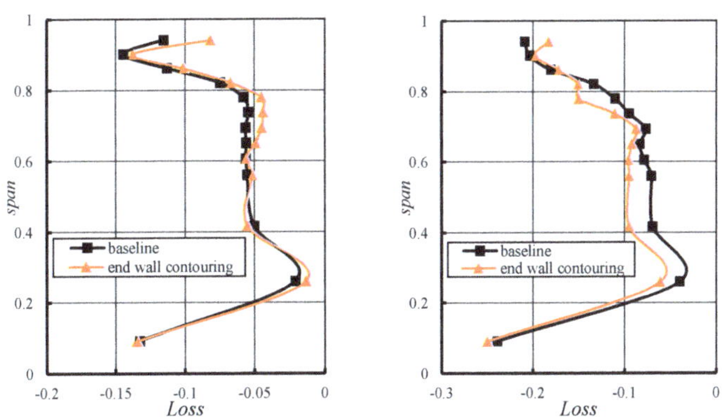

(a) (b)

Figure 15. The circumferential averaged loss coefficient of the stator at 70% design speed. (**a**) $\overline{m}$ = 0.89; (**b**) $\overline{m}$ = 0.77.

At the BM point, the end wall contouring shows a noticeable effect on reducing end wall loss. According to the experimental data in Figure 15a, the total pressure loss coefficient decreases by about 0.01 from 0.7*span* to the blade tip and nearly remains the same in other regions. The reduction of the total pressure loss reaches 0.035 above 0.9*span* and 0.01–0.013 from 0.8*span* to 0.9*span*. The two-dimensional distribution of σ in the outlet plane of the stator is given in Figure 16a,b. This reveals that the low value of σ significantly reduces and moves farther away from the end wall than the baseline stator. This is apparently consistent with the mechanism in the cascade and previous research [4,9].

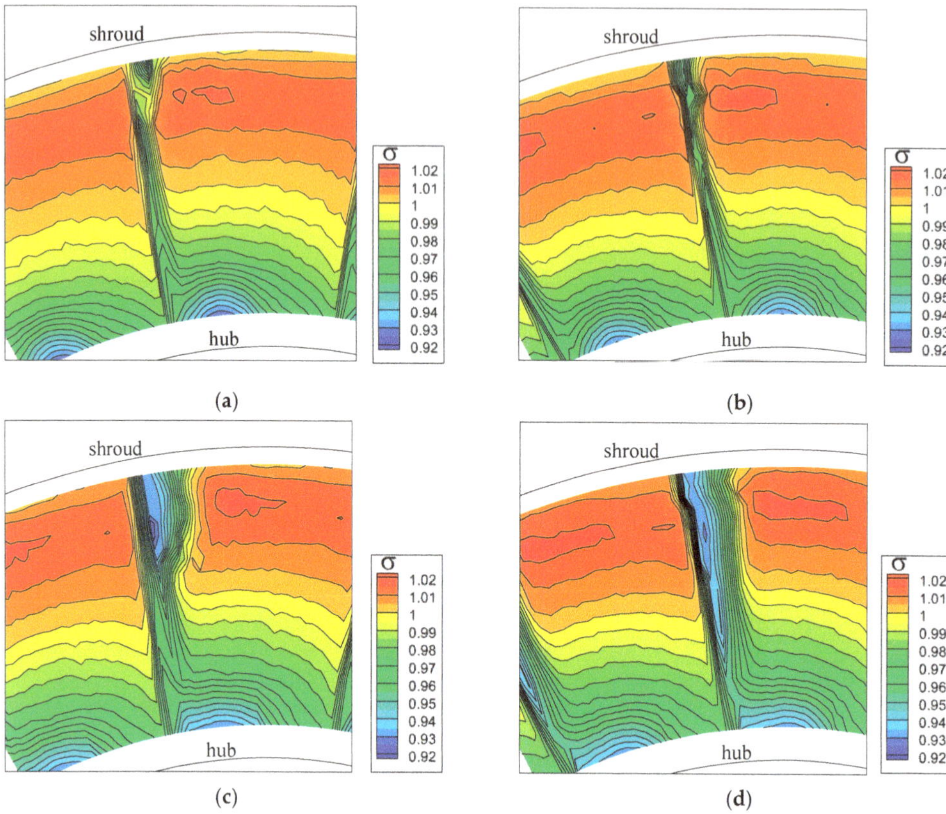

Figure 16. The two-dimensional distribution of total pressure recovery coefficient at the outlet plane of the stator: (**a**) baseline, $\overline{m} = 0.89$; (**b**) end wall contouring, $\overline{m} = 0.89$; (**c**) baseline, $\overline{m} = 0.77$; (**d**) end wall contouring, $\overline{m} = 0.77$.

At the LM point, the effect of end wall contouring on the loss coefficient extends significantly from the blade top to the full-span region. Figure 15b shows that the loss coefficient reduces by 0.01–0.03 above 0.85*span*, increases by 0.01–0.03 from 0.7*span* to 0.85*span*, and increases up to 0.023 from 0.6*span* to the hub. The variation of loss coefficiency from 0.7*span* to the blade tip region should be affected by the end wall contouring. Although part of this region increases the loss coefficient, this still shows more benefit than case c3 in the cascade: the loss coefficient close to the end wall reduces (Figure 15b), and the core region of low σ at 0.75*span* is slightly relieved (Figure 16d). However, the range from 0.6*span* to the hub is beyond the end wall contouring's direct influence. In Figure 16d, the corresponding region shows a heavier wake in the end wall contouring case, indicating a greater blade profile loss than the baseline stator. Thus, the increase of loss below 0.6*span* should be associated with the spanwise variation of the throughflow under the same mass flow rate. It should be the end wall contouring that relieves the blockage due to corner separation on the top side of the blade passage, thereby reducing the mass flow rate of the lower span range. As a result, the lower part of the passage is more throttled than ever, thus generating a higher loss in return.

The above study generally verifies the effect of end wall contouring on the control of corner separation and its contribution to the compressor's performance. However, it is still uncertain whether the control of secondary flow that leads to these improvements is consistent with the design intent shown by the variables in Table 4. This is further verified below with the testing result from the end wall pressure taps.

It is known that near the end wall region, the boundary layer motion should obey the N-S equation, i.e.,

$$\frac{\partial \boldsymbol{V}}{\partial t} + \boldsymbol{V}\nabla\boldsymbol{V} = -\frac{1}{\rho}\nabla p + \tau \tag{11}$$

Suppose the end wall contouring surface has a relatively slight waviness to the scale of the chord length, and the flow is constant; in this case, the N-S equation on the end wall surface before and after the end wall contouring can be subtracted to obtain

$$\boldsymbol{V}_{\text{EWC}}\nabla\boldsymbol{V}_{\text{EWC}} - \boldsymbol{V}_{\text{baseline}}\nabla\boldsymbol{V}_{\text{baseline}} = -\frac{1}{\rho}\nabla(P_{\text{EWC}} - P_{\text{baseline}}) + \boldsymbol{\tau}_{\text{EWC}} - \boldsymbol{\tau}_{\text{baseline}} \tag{12}$$

where the subscripts "EWC" and "baseline" represent the aerodynamic parameters before and after using the end wall contouring, and their subtraction result thus represents the variation of the parameters at the corresponding position on the S1 flow surface. The left side is the variation of the acceleration due to the end wall contouring, which directly changes the secondary flow on the end wall surface. Thus, the variation of pressure gradient on the right side of Equation (12) should be the primary cause for the left-side items, and thus the end wall secondary flow and corner separation. Taking into account the convenience of the testing, here we define a new variable from the static pressure taken by the end wall pressure taps according to

$$\boldsymbol{D} = -\left(\nabla\left(\frac{p}{p_{t0}}\right)_{\text{EWC}} - \nabla\left(\frac{p}{p_{t0}}\right)_{\text{baseline}}\right) \tag{13}$$

where p_{t0} represents the inlet total pressure of the rotor. The variable $\boldsymbol{D}$ represents a quantitative measure of the driving force of the end wall contouring on the end wall secondary flow. The magnitude of $\boldsymbol{D}$ is plotted in Figure 17, and the arrow illustrates its direction.

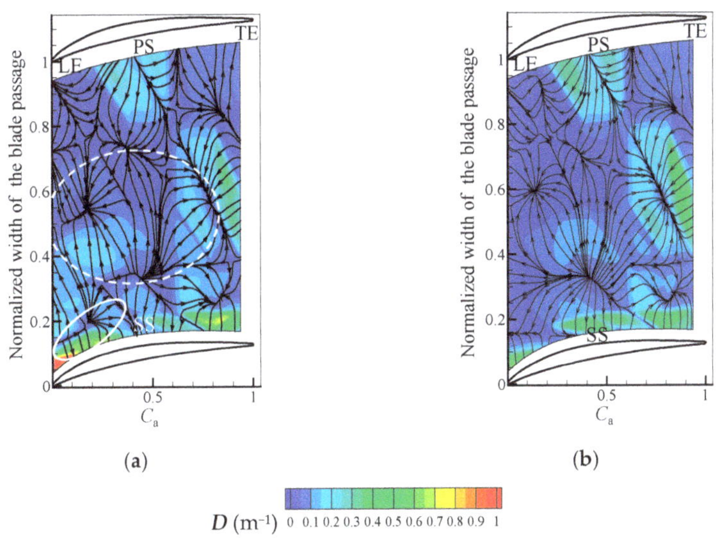

Figure 17. The distribution of $\boldsymbol{D}$ on the end wall surface. (**a**) $\overline{m}$ = 0.89; (**b**) $\overline{m}$ = 0.77.

The contour of the BM point shows clear correspondence with the previously mentioned design intent. From LE to the $0.3C_a$, a high value of $\boldsymbol{D}$ pointing to the SS (marked with a white dashed circle) agrees with the intent of the third localized unit. A wide range of $\boldsymbol{D}$ pointing to the PS (marked with a solid white circle) spreads from the outer SS corner to the middle of the channel. This is associated with the second localized unit. Downstream of $0.4C_a$, the direction of $\boldsymbol{D}$ near the SS corner points to SS from $0.4C_a$ to $0.7C_a$ and then points to the PS direction downstream of $0.7C_a$. The distribution of $\boldsymbol{D}$ here is partially associated with the flow control intent of the first localized unit, except for the region after $0.7C_a$. This is likely related to the downslope surface at the end wall, i.e., the side effect of the localized unit mentioned in Section 2.

Compared to the BM point, the magnitude of $\boldsymbol{D}$ spreading along the SS corner area generally decreases at the LM point. The changes of $\boldsymbol{D}$ with the working condition are not significant. The primary difference lies in the front SS corner, where the direction of $\boldsymbol{D}$ changes to the PS direction. It is possibly related to the reverse flow caused by the separation at the LE corner region, but it can be seen from the figure that the $\boldsymbol{D}$ is not opposed to the streamwise direction. Thus, this shows that end wall contouring somewhat avoids the aggravation of reverse flow caused by the first localized unit. It may likely explain why the end wall contouring, in this case, improves the flow field in the end wall area even at the LM working point.

4.3. Summary of the Effects of the End Wall Contouring

According to the overall performance, the end wall contouring only benefits the right-side operating map of the compressor stage. At the BM point of the compressor, the end wall contouring increases the peak value of σ in the stator, thus making a 10.44% relative reduction of stator loss, which finally increases the overall efficiency by 0.45%. At the LM point, the decrease of σ shows the stator loss increases relatively by 1.95%. Thus, the overall efficiency of the stage decreases by 1.5%.

However, although the efficiency does not increase in all the working conditions, the flow field shows that applying the new end wall contouring method can always improve the corner separation. The tested end wall static pressure shows that the end wall contouring suppresses the end wall secondary flow in the full range but accelerates the cross and climbing flow in the SS corner. The corner separation is thus reduced, and the total pressure loss coefficient in the end wall region reduces locally by 0.01–0.03 for both the operating points.

For the reduction of efficiency at the LM point, the experimental data show it is likely a side effect of suppressing the corner separation. Because the relief of corner separation reduces the local blockage, the throughflow increases in the end wall region and decreases in the main flow. As a result, the main flow region is more "throttled", thus negatively affecting the control of loss. It should be noted that this side effect does not have to increase loss. Just as reported in Hu's research [7], if the main flow is robust, suppressing the corner separation will not increase the loss in the main flow region. Otherwise, it could induce a detrimental effect, as in Reising's research [18], where severe corner separation occurs in the opposite end wall and threatens the stall margin. Considering that even applying an end wall suction of the boundary layer has been reported to increase the loss of main flow [28], the side effect mentioned above should be regarded as a common limitation for all the end wall flow control techniques. The blade row's sensitivity to throttling determines whether or not the loss will increase and how much it will increase. So, for the LM point of the current study, the efficiency reduction reflects that the effect of the end wall contouring was limited by the characteristics of the current compressor stator. The problem may be solved by improving the blade profile.

Above all, the numerical and experimental results indicate that when applying the newly developed end wall contouring method, the intended control of secondary flow in the design space agrees with its actual effect and effectively suppresses the corner separation under multiple inflow conditions. Thus, the results prove its effectiveness. If looking at the empirical end wall contouring method in previously reported studies, the traditional empirical method typically constructs the full area of the end wall surface uniformly through a single analytical formula [2,7,8] or by manual adjustment [9]. In comparison, the newly developed method enables the parametric end wall to control secondary flow in more than one local region under multi-operating conditions. Thus, the control of secondary flow will be more adaptable to different compressors and produce a much closer result to the optimization design, with a relatively small number of design variables. Moreover, in the current new empirical method, the design variables show not only an intuitive effect on the geometry but also a clear relation with the movement of the secondary flow. The new method is thus more convenient for concluding the design experience and shows a possibility to be applied in an optimization process.

5. Conclusions

This study proposes a new end wall contouring method based on the secondary flow control mechanism established in previous research, then verified its effectiveness numerically in a cascade and experimentally in an axial compressor stage. The following conclusions were obtained:

1. The idea of the new end wall contouring method is to define multiple standard surface "units" with particular effects on the end wall secondary flow. Then, we apply a weighted superposition to the units to obtain a comprehensive flow control effect. Compared with the traditional empirical method in the previous research, the design space of the new method enables the parametric end wall to control secondary flow intuitively in more than one local region with a proper amount of design variables, thus showing more advantages.
2. The numerical and experimental research indicates two primary mechanisms of applying the end wall contouring to control the corner separation, i.e., the local acceleration of secondary flow at the SS corner and the full-range deceleration of the secondary flow in the rear passage. The former will accelerate the corresponding secondary climbing flow on the SS, thus mixing with the low-energy separation flow and reducing the separation loss. The latter will suppress the accumulation of low-energy fluid of the boundary layer at the SS corner, thus relieving the local reverse flow. However, the above flow control may also be accompanied by the negative effect of the inverse pressure gradient at the front SS corner, which may seriously deteriorate the corner separation when it is under high incidence.
3. When applied to the stator casing end wall of an axial compressor, although the efficiency only increases on the right part of the operating map, the experiment results show that the intended control of secondary flow agrees with its actual effect and suppresses the corner separation under multiple operating conditions. Its effectiveness is thus verified. The reduction of efficiency at the small mass-flow-rate working points reflects a limitation of the new method and is probably associated with the characteristics of the current compressor stator.

The number of design variables in the newly developed end wall contouring method is relatively small, and all of them show a clear and intuitive effect on the secondary flow. The next step of the research should consider the optimization process. Three-dimensional blading should also be included to overcome any shortages when suppressing the corner separation.

Author Contributions: Funding acquisition, H.Z. and W.C.; Software, Q.L.; Supervision, X.L.; Writing—original draft, F.Y. All authors have read and agreed to the published version of the manuscript.

Funding: The work is funded by the National Natural Science Foundation of China (Grant No. 51906027, 52176036 and 52076179), Doctoral Initiation Research Funds of Liaoning province (Grant No. 2019-BS-027) and the Fundamental Research Funds for the Central Universities (Grant No. 3132022124).

Conflicts of Interest: The authors declare that they have no known competing financial interests or personal relationships that could have appeared to influence the work reported in this paper.

References

1. Rose, M.G. Non-Axisymmetric Endwall Profiling in the HP NGV's of an Axial Flow Gas Turbine. In Proceedings of the ASME Turbo Expo, The Hague, The Netherlands, 13–16 June 1994. 94-GT-249.
2. Meng, T.; Yang, G.; Zhou, L.; Ji, L. Full blended blade and endwall design of a compressor cascade. *Chin. J. Aeronaut.* **2021**, *34*, 79–93. [CrossRef]
3. Ma, Y.; Teng, J.; Zhu, M.; Qiang, X. Non-axisymmetric Endwall Contouring in a Linear Compressor Cascade. In Proceedings of the ASME Turbo Expo 2020, London, UK, 21–25 September 2020. GT2020-14366.
4. Li, X.; Chu, W.; Wu, Y.; Zhang, H.; Spence, S. Effective end wall profiling rules for a highly loaded compressor cascade. *Proc. Inst. Mech. Eng. A J. Power Energy* **2016**, *230*, 535–553. [CrossRef]
5. Liu, X.; Jin, D.; Gui, X.; Liu, X.; Guo, H. Effect of Solidity on Non-Axisymmetric Endwall Contouring Performance in Compressor Linear Cascades. In Proceedings of the ASME Turbo Expo, Oslo, Norway, 11–15 June 2018. GT2018-76247.
6. Ma, Y.; Teng, J.; Zhu, M.; Qiang, X. Influence of the inlet boundary layer on non-axisymmetric endwall contouring effects in a linear compressor cascade. *Proc. Inst. Mech. Eng. C J. Mech. Eng. Sci.* **2022**. [CrossRef]
7. Hu, S.; Lu, X.; Zhang, H.; Zhu, J.; Xu, Q. Numerical Investigation of a High-subsonic Axial-flow Compressor Rotor with Non-axisymmetric Hub Endwall. *J. Therm. Sci.* **2010**, *19*, 14–20. [CrossRef]
8. Cao, Z.; Gao, X.; Liu, B. Control mechanisms of endwall profiling and its comparison with bowed blading on flow field and performance of a highly-loaded compressor cascade. *Aerosp. Sci. Technol.* **2019**, *95*, 105472. [CrossRef]
9. Harvey, N.W. Some Effects of Non-Axisymmetric End wall Profiling on Axial Flow Compressor Aerodynamics. Part I: Linear Cascade Investigation. In Proceedings of the ASME Turbo Expo, Berlin, Germany, 9–13 June 2008. GT2008-50990.
10. Reising, S.; Schiffer, H. Non-axisymmetric end wall profiling in transonic compressors. Part II: Design study of a transonic compressor rotor using nonaxisymmetric end walls optimization strategies and performance. In Proceedings of the ASME Turbo Expo, Orlando, FL, USA, 8–12 June 2009. GT2009-59134.
11. Zhang, X.; Lu, X.; Zhu, J. Performance Improvements of a Subsonic Axial-Flow Compressor by Means of a Non-Axisymmetric Stator Hub End-Wall. *J. Therm. Sci.* **2013**, *22*, 539–546. [CrossRef]
12. Dorfner, C.; Hergt, A.; Nicke, E.; Moenig, R. Advanced Nonaxisymmetric End wall Contouring for Axial Compressors by Generating an Aerodynamic Separator- Part I: Principal Cascade Design and Compressor Application. *J. Turbomach.* **2011**, *133*, 021026. [CrossRef]
13. Hergt, A.; Dorfner, C.; Steinert, W.; Nicke, E.; Schreiber, H. Advanced Nonaxisymmetric End wall Contouring for Axial Compressors by Generating an Aerodynamic Separator- Part II: Experimental and Numerical Cascade Investigation. *J. Turbomach.* **2011**, *133*, 021027. [CrossRef]
14. Reutter, O.; Hemmert-Pottmann, S.; Hergt, A.; Nicke, E. Endwall contouring and fillet design for reducing losses and homogenizing the outflow of a compressor cascade. In Proceedings of the ASME Turbo Expo, Düsseldorf, Germany, 16–20 June 2017. GT2014-25277.
15. Huang, S.; Yang, C.; Li, Z.; Han, G.; Zhao, S.; Lu, X. Effect of Non-Axisymmetric End Wall on a Highly Loaded Compressor Cascade in Multi-Conditions. *J. Therm. Sci.* **2021**, *30*, 1363–1375. [CrossRef]
16. Harvey, N.W.; Offord, T.P. Some Effects of Non-Axisymmetric End wall Profiling on Axial Flow Compressor Aerodynamics. Part II: Multi-Stage HPC CFD Study. In Proceedings of the ASME Turbo Expo, Berlin, Germany, 9–13 June 2008. GT2008-50991.
17. Varpe, M.K.; Pradeep, A.M. Benefits of nonaxisymmetric endwall contouring in a compressor cascade with a tip clearance. *J. Fluids Eng.* **2015**, *137*, 051101. [CrossRef]
18. Reising, S.; Schiffer, H. Non-axisymmetric end wall profiling in transonic compressors. Part I: Improving the static pressure recovery at off-design conditions by sequential hub and shroud end wall profiling. In Proceedings of the ASME Turbo Expo, Orlando, FL, USA, 8–12 June 2009. GT2009-59133.
19. Lepot, I.; Mengistu, T.; Hiernaux, S. Highly Loaded LPC Blade and Non Axisymmetric Hub Profiling Optimization for Enhanced Efficiency and Stability. In Proceedings of the ASME Turbo Expo, Vancouver, BC, Canada, 6–10 June 2011. GT2011-46261.
20. Li, X.; Chu, W.; Wu, Y. Numerical investigation of inlet boundary layer skew in axial-flow compressor cascade and the corresponding non-axisymmetric end wall profiling. *Proc. Inst. Mech. Eng. A J. Power Energy* **2014**, *228*, 638–656. [CrossRef]
21. Ji, L.; Tian, Y.; Li, W.; Yi, W.; Wen, Q. Numerical Studies on Improving Performance of Rotor-67 by Blended Blade and Endwall Technique. In Proceedings of the ASME Turbo Expo, Copenhagen, Denmark, 11–15 June 2012. GT2012-68535.
22. Mahmood, S.; Turner, M.; Siddappaji, K. Flow Characteristics of an Optimized Axial Compressor Rotor Using Smooth Design Parameters. In Proceedings of the ASME Turbo Expo, Seoul, Korea, 13–17 June 2016. GT2016-57028.
23. Cheng, J.; Chen, J.; Xiang, H. A Surface Parametric Control and Global Optimization Method for Axial Flow Compressor Blades. *Chin. J. Aeronaut.* **2019**, *32*, 1618–1634. [CrossRef]
24. Sun, S.; Chen, S.; Liu, W.; Gong, Y.; Wang, S. Effect of Axisymmetric Endwall Contouring on the High-load Low-reaction Transonic Compressor Rotor with a Substantial Meridian Contraction. *Aerosp. Sci. Technol.* **2018**, *81*, 78–87. [CrossRef]
25. Brennan, G.; Harvey, N.W.; Rose, M.G.; Fomison, N.; Taylor, M.D. Improving the Efficiency of the Trent 500-HP Turbine Using Nonaxisymmetric End Walls—Part I: Turbine Design. *ASME J. Turbomach.* **2003**, *125*, 497–504. [CrossRef]
26. Zhang, Y.; Mahallati, A.; Benner, M. Experimental and numerical investigation of corner stall in a highly-loaded compres-sor cascade. In Proceedings of the ASME Turbo Expo, Düsseldorf, Germany, 16–20 June 2014. GT2014-27204.

27. Akcayoz, E.; Vo, H.D.; Mahallati, A. Controlling Corner Stall Separation with Plasma Actuators in a Compressor Cascade. In Proceedings of the ASME Turbo Expo, Montreal, QC, Canada, 15–19 June 2015. GT2015-43404.
28. Lu, H.; Wang, X.; Guo, S.; Huang, Y.; Zheng, Y.; Zhang, H.; Zhong, J. The Numerical Investigation of Asymmetric Boundary Layer Suction in High-load Compressor Cascades. *J. Eng. Thermophys.* **2018**, *39*, 977–984.

applied sciences

MDPI

Article

Numerical Simulation of Random Cavitation Suppression Based on Variable NACA Airfoils

Weidong Shi [1,2,*], Zhouhao Shi [1], Zhanshan Xie [1,2,*], Qinghong Zhang [1], Yongfei Yang [1,2] and Linwei Tan [1,2]

1 School of Mechanical Engineering, Nantong University, Nantong 226019, China; shizhouhao@126.com (Z.S.); zhangqinghong2021@126.com (Q.Z.); yyf2020@ntu.edu.cn (Y.Y.); tanlinwei@ntu.edu.cn (L.T.)
2 Institute of Fluid Machinery and Marine Engineering Equipment, Nantong 226019, China
* Correspondence: wdshintu@163.com (W.S.); xiezs@ntu.edu.cn (Z.X.)

Abstract: In order to suppress the cavitation of an airfoil under random operating conditions, a deformable covering was constructed in the cavitation prone area of the NACA0012 airfoil. By sensing the pressure difference between the inner and outer sides of the airfoil, the covering of the airfoil can be changed adaptively to meet the requirement of suppressing random cavitation of the airfoil. The simulation results show that the cavitation influence range of the airfoil with a shape memory alloy covering can be reduced by more than 70%, and the cavitation is well reduced and suppressed. Moreover, the backflow near the wall of the airfoil was reduced under random working conditions. When the maximum bulge deformation of the covering was between 3–6 mm, the airfoil produced a cavitation range only on the covering surface of the airfoil, and there was no cavitation erosion on other parts. This method with locally variable airfoil to suppress cavitation provides a good reference value for other hydraulic machinery to suppress cavitation.

Keywords: NACA hydrofoil; cavitation suppression; random operating condition

Citation: Shi, W.; Shi, Z.; Xie, Z.; Zhang, Q.; Yang, Y.; Tan, L. Numerical Simulation of Random Cavitation Suppression Based on Variable NACA Airfoils. *Appl. Sci.* **2021**, *11*, 11618. https://doi.org/10.3390/app112411618

Academic Editor: Jesús María Blanco

Received: 29 September 2021
Accepted: 1 December 2021
Published: 7 December 2021

1. Introduction

Cavitation flow is a complex multiphase turbulence, including phase transition, flow separation and multi-scale vortex structure. Cavitation often occurs at the low-pressure surface of a flow passage component (such as the suction surface of a pump or the lifting surface of marine propeller [1–3]). It is the main cause of performance deterioration, vibration, noise, and even surface erosion of underwater equipment [4–6].

In recent years, with the rapid development of computers, a numerical simulation method has provided important assistance for complex experiments, which has become the key technology for researchers to use to analyze flow fields. CFD technology has become very common to study the inherent laws and mechanism of fluid movement. V Ryzhenkov [7] analyzed the time-averaged velocity field and fluctuations of annular jet flow, which proved that the re-circulation region was accompanied by low-frequency oscillations. Wei Zhang [8] carried out numerical simulation of hydrofoil flow; he used a finite mass transfer model of cavitation to capture the condensation shock wave, and the three-dimensional process of the shock wave could be clearly observed in the simulation. Bhatt and Mahesh [9] used compressible large eddy simulation (LES) to simulate cavitation on the same geometry of Ganesh et al. [10]. They found that the results were in good agreement with the experimental results, which confirmed the existence of bubble shock propagation and the effect of compressibility. Ku Garam [11] developed a numerical method to research the tip vortex cavitation and flow noise.

In terms of cavitation inhibition, many scholars inhibit cavitation by adding inducers [12], modifying blade shape, size [13,14], and other special methods, such as blade gap jet [15–17], protruding modification [18,19], microtexture surface modification [20], and so on. In addition, scholars have obtained quantitative data about cavitation and turbulence through experiments and high-speed camera observations [21–25]. For example, Jun-ye

Li [26] researched the effects of the outlet pressure, the operating temperature, and the installation angle of the two cages on the cavitation and found out the best method to suppress cavitation. Shin Suyong [27] researched the physical characteristics of cavitation initiation, the rotational motion of the tip vortex cavitation flow, and the characteristics of induced noise during vortex development. Choi [28] observed the whole process of cavitation flow, vortex cavitation, tip leakage vortex cavitation, and void formation in a turbine pump inducer. On the basis of the study of blade shape control, and referring to the blade control technology studied by predecessors, a novel NACA airfoil with a covering is proposed to improve the cavitation performance. This new cavitation method can effectively block the re-entry jet to improve the flow field around the cover, and it restrain the development of the low-pressure area. Additionally, it is a good reference for other hydraulic machinery to suppress cavitation.

2. Physical and Mathematical Model

2.1. Geometrical Model and Mesh

Figure 1 shows the modified covering airfoil based on NACA0012. The covering is made of Fe-Mn-Si shape memory alloy, which is set at a cavitation-prone position. Additionally, the constant pressure fluid ($p_{in} = 1$ bar) is passed into the covering. Under normal working conditions, the internal and external pressure difference does not deform the memory metal within a certain range. However, cavitation of the airfoil occurs under abnormal operating conditions, and under the condition of increasing pressure difference between the inner and outer sides, the center of the memory alloy covering will deform and eventually form an eggshell bulge and cavitation will suppressed by impeding backflow [29].

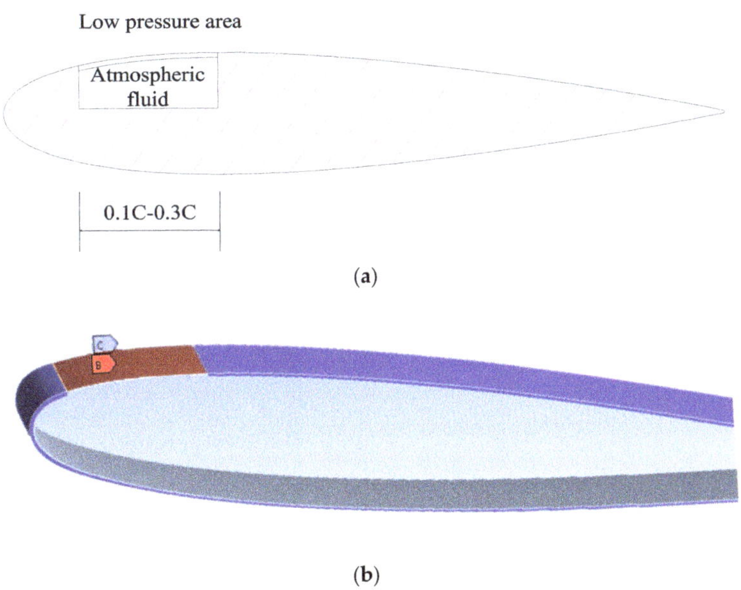

Figure 1. Schematic diagram of improved covering airfoil. (**a**) The cross-sectional view of the airfoil and the variable area of the airfoil. (**b**) 3D diagram of a variable airfoil.

According to the literature and the research results of other experts and scholars on multiphase flow models, the Euler–Euler model allows the simulation of multiple independent and interacting phases; that is, the interaction between the cavitation bubble and the liquid can be simulated more accurately. Therefore, the Euler model is adopted as a multiphase flow model for simulation in this manuscript.

2.2. Cavitation Model and Turbulence Model

In view of the research results [30,31], for the commonly used Smagorinsky model of the LES method and the standard SST model based on the RANS method, the cavitation model of the viscosity-modified SST model can accurately predict the cavitation shedding frequency and simulates the return jet at the wake of the hydrofoil (which is the main cause of cavitation shedding). For the turbulence model, it is necessary to calculate the dense boundary layer with high Reynolds number, and the k-ε turbulence model is suitable for this kind of working condition. Therefore, the SST cavitation model and the k-ε turbulence model are used for numerical simulation. The vaporization rate and liquefaction rate derived from the Rayleigh–Plesset equation in the Schnerr–Sauer model are shown in Formulas (1) and (2), respectively, and the volume fraction of gas phase (α_v) is shown in Formula (3).

$$\dot{m}^- = \frac{\rho_1 \rho_v}{\rho_m} \frac{3\alpha_v(1-\alpha_v)}{R_B} \sqrt{\frac{2}{3} \frac{(p_v - p)}{\rho_m}} \tag{1}$$

$$\dot{m}^+ = -\frac{\rho_1 \rho_v}{\rho_m} \frac{3\alpha_v(1-\alpha_v)}{R_B} \sqrt{\frac{2}{3} \frac{(p - p_v)}{\rho_m}} \tag{2}$$

$$\alpha_v = n_0 \frac{4}{3} \pi R^3 / \left(n_0 \frac{4}{3} \pi R^3 + 1 \right) \tag{3}$$

Furthermore, we modify the turbulent viscosity coefficient in the SST k-ε model, and its expression is shown in Formulas (4)–(6).

$$u_t = f(p) C_\omega \frac{k}{\omega} \tag{4}$$

$$C_w = \frac{1}{max\left[\frac{1}{a^*}, \frac{SF}{a_1 w}\right]} \tag{5}$$

$$f(\rho) = \rho_v + \frac{(\rho_m - \rho_v)^n}{(\rho_l - \rho_m)^{n-1}} \tag{6}$$

In the above formula, R_B is the bubble radius; n_0 is the density of gas nuclei; p, p_v are local pressure and saturated vapor pressure, respectively; ρ_m, ρ_1, ρ_v are mixed phase density, liquid phase density and gas phase density, respectively; u_t is turbulent viscosity; and k and ω are turbulent kinetic energy and dissipation rate.

3. Settings for Calculation

3.1. Flow Field Calculation Settings

In most of the simulation studies by experts and scholars, the 2D flow mode is mostly used in the researches of NACA airfoils. Compared with 3D simulations, the computational resources required are much lower and the calculation time can be saved. However, there is a significant difference between 2D and 3D simulations for most flow conditions [32]. In the real world, there are also some special turbulences such as flow around a cylinder, a three-dimensional vortex, and so on. There are significant inherent differences between 2D and 3D simulation predictions. After comprehensive consideration, this simulation is based on Ansys 2020R2 for 3D modeling, and the SST cavitation model and k-ε turbulence model are used to simulate the flow field.

The calculation domain is shown in Figure 2. The airfoil is located at the center of the upper and lower boundaries, with the head of the airfoil being 3C away from the inlet edge. The upper and lower boundaries are 2.5C, respectively, and the outlet edge is 6C away from the trailing edge. The geometric parameters of airfoil are chord length C = 1 m and incoming flow angle $\alpha = 6°$. The cavitation number is a common standard to measure the severity of airfoil cavitation, and it is also commonly used to identify different cavitation conditions.

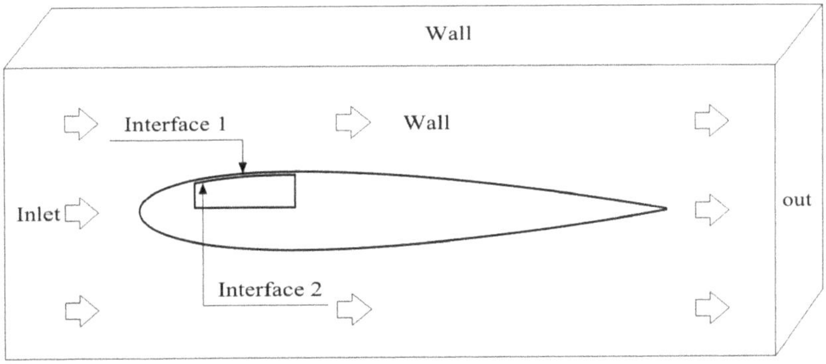

Figure 2. NACA airfoil calculation domain diagram.

The quality and quantity of the mesh have a significant influence on the calculation results, so the structured mesh was used in the hydrofoil calculation domain, and the mesh around the foil was encrypted. In addition, we carried out a grid-independent analysis and evaluated the cavitation number of airfoils according to Formula (7). The results are shown in Table 1.

$$\sigma = \frac{p_\infty - p_v}{0.5\rho U^2} \quad (7)$$

Table 1. Mesh independence test.

Case	Mesh Number	Inlet Pressure/Pa	σ
A1	551,300	134,529	0.656
A2	1,380,200	135,378	0.661
A3	2,741,200	135,813	0.662
A4	11,057,200	136,033	0.663

In the formula, p_∞ is the inlet pressure, saturated vapor pressure p_v = 3540 P_a, U is the velocity of incoming flow at infinity. Therefore, when the boundary conditions are not changed, the cavitation number will not change, and it is used to evaluate the mesh independence.

It can be seen from Table 1 that the difference between the cavitation number σ and the number from Case A2 to Case A4 is less than 0.2%. Considering the time limitation, the Case A2 mesh is chosen for simulation.

3.2. Solid Field Calculation Setting

The fluid–structure coupling surface is set on the front and back of the covering, and the flow field data is introduced into the solid field by pressure (as shown Figure 3). Because the Fe-Mn-Si metal is a metallic material with memory properties, it is necessary to define the material properties in Ansys2020R2. Considering the need for a certain amount of tensile strength and ductility, silicone rubber is used as the base material because it has similar properties, and some parameters are modified: elastic modulus $E = 2 \times 10^5$ MPa, Poisson's ratio $\sigma = 0.48$, and tensile strength $R_m = 5.02$ MPa. In addition, because the memory metal has good toughness and does not fold easily, the airfoil tail was rounded. Additionally, the maximum deformation of the covering can be changed by the internal and external pressure of the covering and the thickness of the covering, when the pressure difference is 100 Mpa and the thickness is 1 mm as the basic parameter of deformation, the deformation of the memory metal is shown in Figure 3. Then, the internal pressure is adjusted to produce airfoil with different degrees of bulge. The airfoil models with different degrees of bulge were derived and calculated for the same operating conditions and compared with the original flow field.

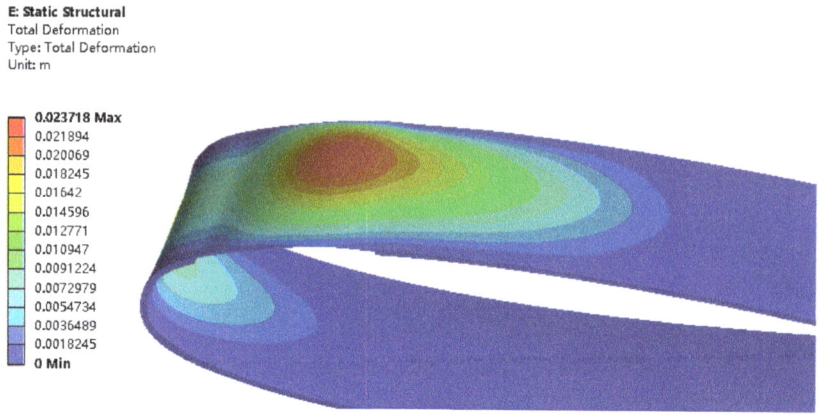

Figure 3. Adaptive change nephogram of airfoil structure in cavitation area.

4. Results and Discussion

4.1. Cavitation Nephogram Discussion

Figure 4a shows the cavitation nephogram of original airfoil, and Figure 4b,d shows the cavitation nephogram of the airfoil with convexity of 1 mm, 3 mm, and 6 mm, respectively. The cavitation bubbles of the original airfoil are tight against the upper surface and extend in the direction of the oncoming flow, which accounts for about 2/3 of the chord length of the airfoil, and the upper surface of the airfoil is heavily cavitated. Compared with the original airfoil, it is found that the cavitation range of airfoil with maximum variable of 1 mm is obviously reduced, the main body of the cavitation area is inverted trapezoid, and there is partial vortex cavitation at the edge of the airfoil, which is mainly caused by the initial protrusion of the covering in the middle of the deformation. The cavitation area of airfoil with convex degree of 3 mm is further reduced to about 1/4 of the upper surface. The cavitation deviation caused by the height difference still exists, but it decreases with the increase of convex degree. As for the airfoil with a maximum variable of 6 mm, cavitation occurs only at the raised position of the covering, and cavitation and the low-pressure area are no longer generated elsewhere on the upper surface. Moreover, the cavitation scale of the covering is significantly reduced, which indicates that the bulge of the covering plays an effective role in inhibiting the development of cavitation.

4.2. Pressure Graph Discussion

The pressure in the flow field of the airfoil directly determines the cavitation characteristics of the airfoil. In this section, the coordinates and pressure points of one hundred points on the upper surface are taken and compared. Figure 5 shows the pressure curves for upper surface of the original airfoil and the airfoil with different protuberances. The low-pressure area of the original airfoil starts from 0.03C (at the head of the airfoil) to 0.6C (in the middle of the airfoil). This part of the low-pressure area is the cause of cavitation generation and development. Additionally, it is consistent with the cavitation nephogram. The total length of the low-pressure area of the airfoil with a maximum variable of 3 mm is about 0.4C, which is 30% smaller than that of the control group. Additionally, the pressure curve produces some fluctuations at 0.05C, which are caused by the sheet cavitation in front of the bulge. After the degree of convexity of the covering from 1 mm to 3 mm, it obstructs the flow and forms a small high-pressure zone at 0.1C, and the total length of the low-pressure area is about 0.25C, which is about 61% less than the control group. For the airfoil with maximum variable of 6 mm, the fluctuations still exists at 0.05C, but due to the increase in the bulge, the pressure does not reach the critical pressure of cavitation, so the sheet cavitation before the bulge no longer occurs. It can be observed from the figure that the low pressure area is distributed in 0.1–0.2C, the cavitation range is reduced by more than 80%, and it is completely within the covering range, so the cavitation is far away from the wall and the cavitation collapse will not cause erosion damage to the near wall of the airfoil.

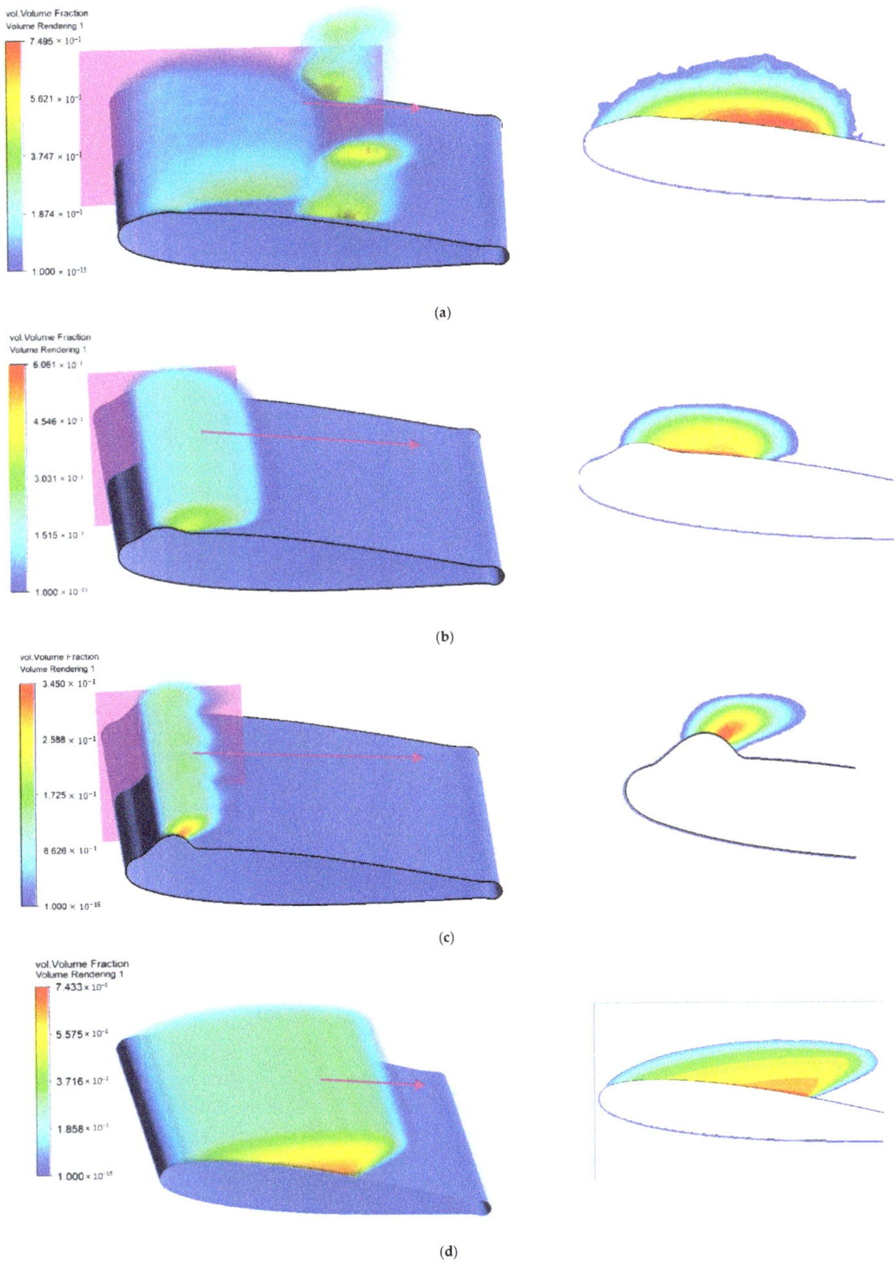

Figure 4. Effect of local deformation on airfoil cavitation suppression. (**a**) Cavitation nephogram of airfoil with maximum variable of 1 mm. (**b**) Cavitation nephogram of airfoil with maximum variable of 3 mm. (**c**) Cavitation nephogram of airfoil with maximum variable of 6 mm. (**d**) Cavitation nephogram of original airfoil.

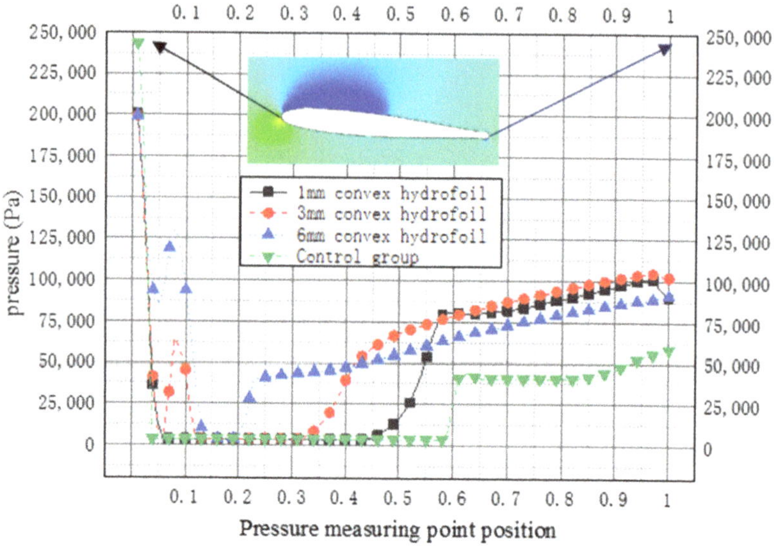

Figure 5. Pressure graph.

4.3. Flow Field and Re-Entrant Jet Discussion

The re-entrant jet is similar to the reflux in boundary layer separation. According to the experimental determination of reference [23], it was found that the frequency of the re-entrant jet was essentially the same as the volume change frequency of cavitation, which caused the cavitation bubbles at the back of the cavitation area to burst and collapse, eventually forming cloud cavitation. Therefore, this section analyzes the flow field and the return jet, which intercepted the velocity vector diagram of the original airfoil and the airfoil with different protruding degrees on the upper surface of the zonal 1.5C section, as shown in Figure 6a. It is obvious that there is a re-entrant jet in the original airfoil, which leads to the cavitation shedding and cloud cavitation phenomena. It can be seen that the re-entry jet is still present on the upper surface of the airfoil with a maximum variable of 1 mm. In addition, the three-dimensional flow field at 0.05C of the airfoil with the degree of bulge of 1 mm is found to be intensified, and due to the bulge in the middle, the thickness of the bulge on both sides is slightly lower than the sides, the fluid in the middle will flow to both sides, and the flow on both sides will be greater than the flow rate in the middle. This causes the fluid on both sides to flow to the middle and continuously develop to the front of the airfoil and eventually form a vortex (as shown in Figure 6b); the vortex cavitation at the edge of the airfoil (show in Figure 4a) is also caused by this phenomenon.

With the increase in the degree of bulge, the re-entrant jet decreases gradually, and when the degree of bulge is 3 mm, the high-speed fluid is blocked, and there is no reverse flow phenomenon (as shown in Figure 7a). When the degree of bulge reaches 6 mm, reverse flow is found again near the bulge (as shown in Figure 7b). This is due to the fact that the high-speed water flow in front of the bulge is mostly blocked, and the backflow is completely blocked by the bulge, so cavitation does not occur.

To sum up, from the point of view of cavitation suppression, the blocking effect of memory metal covering on the development of cavitation is significant. However, the design purpose of airfoil is to provide lift, so considering cavitation alone will lead to the increase in airfoil resistance, and the lift-drag ratio and cavitation resistance should be considered to ensure the mechanical properties of the airfoil. The maximum bulge deformation of the covering is controlled between 3 and 6 mm by adjusting the internal fluid pressure and thickness, which can hinder the development of cavitation under abnormal operating conditions, and it reduce the erosion damage to the wall by cavitation, and it ensures that the normal operation of machinery does not produce negative effects such as vibration and noise. After operating conditions are improved, the memory metal is restored by heating and other methods to ensure the normal operation of machinery.

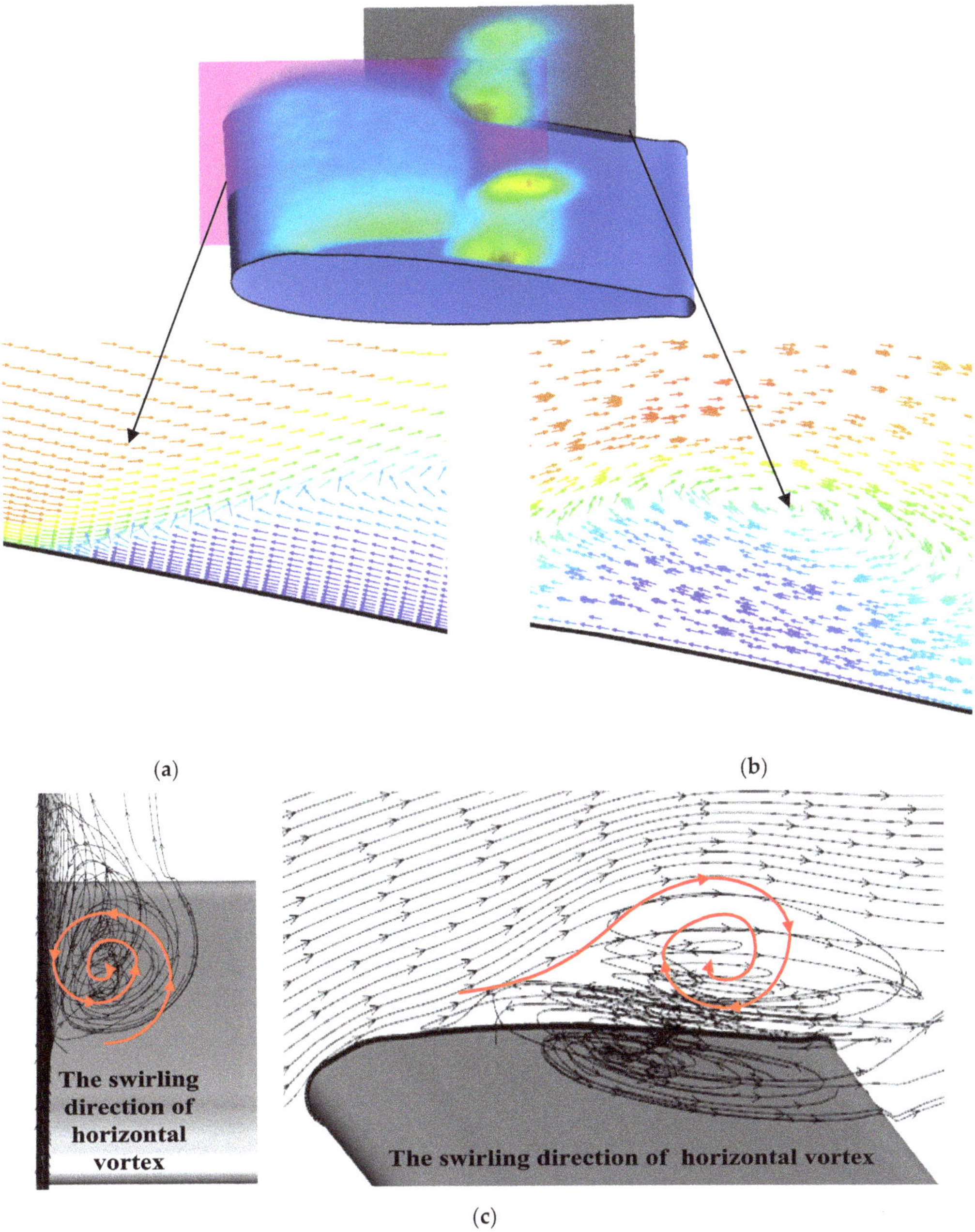

Figure 6. Flow field on the upper surface of airfoil with maximum variable of 1 mm. (**a**) Velocity vector diagram of airfoil with maximum variable of 1 mm at the z = 1.5C section. (**b**) Velocity vector diagram of airfoil with maximum variable of 1 mm at the z = 0.05C section. (**c**) Streamline diagram.

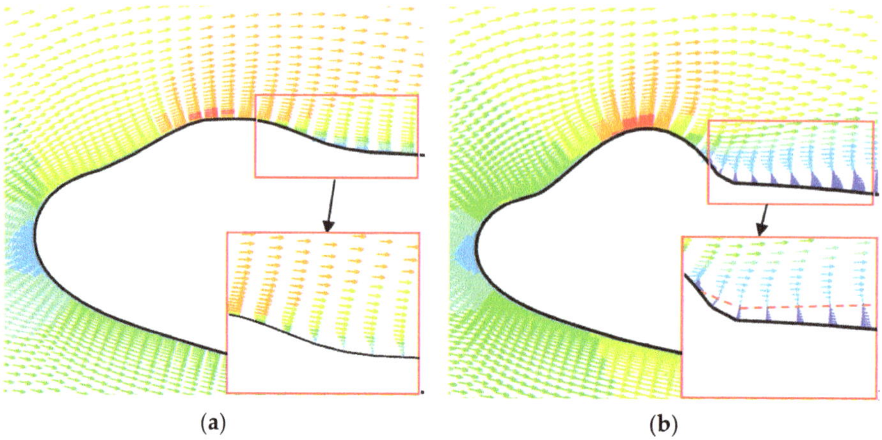

Figure 7. Velocity vector diagram on the upper surface of airfoil. (**a**) Velocity vector diagram of airfoil with maximum variable of 3 mm at the z = 1.5C section. (**b**) Velocity vector diagram of airfoil with maximum variable of 6 mm at the z = 1.5C section.

5. Conclusions

(1) The NACA airfoil with deformable covering effectively hinders the re-entrant jet, and it improves the flow field around the covering and suppresses the development of the low-pressure area.

(2) With the increase in the deformation degree of the deformable airfoil, the cavitation occurrence area is concentrated to the surface of the covering, and the influence range of different degrees of deformation on the cavitation is reduced by 30% to 80%, and the degree of cavitation is reduced by 70%, indicating that this method effectively suppresses the development of cavitation.

(3) The optimal maximum deformation of the deformable covering is between 3 mm and 6 mm, which can hinder the development of cavitation and reduce the erosion of cavitation on the wall under abnormal working conditions, ensuring that the normal operation of machinery does not produce negative effects such as vibration and noise.

Author Contributions: Conceptualization, W.S. and Z.X.; methodology, Z.S.; software, Z.S.; validation, W.S., Z.X. and Y.Y.; formal analysis, Q.Z.; investigation, Q.Z.; resources, W.S.; data curation, L.T.; writing—original draft preparation, Z.S.; writing—review and editing, Z.S.; visualization, Q.Z.; supervision, Z.X.; project administration, L.T.; funding acquisition, W.S. All authors have read and agreed to the published version of the manuscript.

Funding: This research was funded by the National Key Research and Development Project of China (no. 2019YFB 2005300), National High-Tech Ship Scientific Research Project of China (no. MIIT [2019] 360), National Natural Science Foundation of China (No. 51979138), National Natural Science Foundation of China (No. 52109106), General project of natural science research in colleges and universities of Jiangsu Province (No. 19KJB470029), and Jiangsu Water Conservancy Science and Technology Project (No. 2019038).

Data Availability Statement: Data on the analysis and reporting results during the study can be obtained by contacting the authors.

Conflicts of Interest: The authors declare no conflict of interest.

References

1. Lin, P.; Hu, D.; Lin, Z.J.; Liu, M.Q.; Tang, C.L.; Wang, S. The mechanism of joint effects of axial-flow pump cavitation and sediment wear. *Adv. Mech. Eng.* **2020**, *12*, 1687814020923066. [CrossRef]
2. Xie, Z.; Shi, W.; Zheng, Y.; Tan, L.; Tian, Q.; Cao, Y.; Ge, X. Simulation investigation on impact damage characteristics of metal plate by cavitating bubble micro-jet water hammer. *Eng. Fail. Anal.* **2020**, *115*, 783–793. [CrossRef]

3. Wang, G.; Wu, Q.; Huang, B. Dynamics of cavitation–structure interaction. *Acta Mech. Sin.* **2017**, *33*, 685–708. [CrossRef]
4. Shi, Z.; Xie, Z.; Shi, W.; Zhang, Q.; Tan, L. Numerical Investigation on Cavitation Suppression of Microchannel over a NACA0012 Hydrofoil. *Shock. Vib.* **2021**, *2021*. [CrossRef]
5. Huang, B.; Zhao, Y.; Wang, G. Large Eddy Simulation of turbulent vortex-cavitation interactions in transient sheet/cloud cavitating flows. *Comput. Fluids* **2014**, *92*, 113–124. [CrossRef]
6. Xie, Z.; Shi, W.; Tian, Q.; Zheng, Y.; Tan, L. Fatigue Life Assessment and Damage Investigation of Centrifugal Pump Runner. *Eng. Fail. Anal.* **2021**, *124*, 105256. [CrossRef]
7. Ryzhenkov, V.; Mullyadzhanov, R. Direct numerical simulations of the turbulent annular jet with different diameter ratio. *J. Phys. Conf. Ser.* **2018**, *1105*, 403–414. [CrossRef]
8. Zhang, W.; Zhu, B.; Wang, Y.; Xu, H. Numerical simulation of condensation shock in partial cavitating flow on a hydrofoil. *J. Hydrodyn.* **2020**, *32*, 128–136. [CrossRef]
9. Bhatt, M.; Mahesh, K. Numerical investigation of partial cavitation regimes over a wedge using large eddy simulation. *Int. J. Multiph. Flow* **2020**, *122*, 103155. [CrossRef]
10. Ganesh, H.; Mäkiharju, S.A.; Ceccio, S.L. Bubbly shock propagation as a mechanism for sheet-to-cloud transition of partial cavities. *J. Fluid Mech.* **2016**, *802*, 37–78. [CrossRef]
11. Ku, G.; Cheong, C.; Seol, H. Numerical Investigation of Tip-Vortex Cavitation Noise of an Elliptic Wing Using Coupled Eulerian-Lagrangian Approaches. *Appl. Sci.* **2020**, *10*, 5897. [CrossRef]
12. Huan, Y.Y.; Liu, Y.Y.; Li, X.J.; Zhu, Z.C.; Qu, J.T.; Zhe, L.; Han, A.D. Experimental and numerical investigations of cavitation evolution in a high-speed centrifugal pump with inducer. *J. Hydrodyn.* **2021**, *33*, 140–149. [CrossRef]
13. Zhao, W. Elementary analysis of centrifugal pump impeller destruction due to cavitation and sand abrasion. *J. Gansu Univ. Technol.* **2000**, *1*, 58–62. (In Chinese)
14. Yang, C.; Suo, Y.; Zhu, W.; Jiang, Z.; Lao, D. Numerical Investigation of the Internal Flow Field and Redesign of a 30° Backswept Impeller. *Power Eng.* **2000**, *1*, 580–584. (In Chinese)
15. Zhang, R.; Yun, L. Investigation on the effect of slot pulse jet on centrifugal pump performance. Turbomachinery Society of Japan, Korean Society for Fluid Machinery, Chinese Society of Engineering Thermophysics. *Int. J. Fluid Mach. Syst.* **2018**, *11*, 139–145. [CrossRef]
16. Shi, W.; Li, L.; Xu, R.; Wang, B.; Tan, L.; Zhou, L. Effect of blade slotting on radial force of single vane centrifugal pump. *J. Drain. Irrig. Mach. Eng. (JDIME)* **2020**, *38*, 865–870, 890. (In Chinese)
17. Zhao, W.; Lu, J.; Zhao, F. Research on cavitation control of centrifugal pump based on slot jet principle. *J. Zhejiang Univ. (Eng. Ed.)* **2020**, *54*, 1785–1794.
18. Wu, W.; Xiong, Y. A Reshaping Method for Anti-cavitating Hydrofoil Design. *J. Shanghai Jiaotong Univ.* **2013**, *47*, 878–883, 888. (In Chinese)
19. Wu, W.; Xiong, Y.; Qi, W. Cavitation suppression based on airfoil modification. *Res. Chin. Ships* **2012**, *7*, 36–40, 45. (In Chinese)
20. Gonzalez-Avila, S.R.; Nguyen, D.M.; Arunachalam, S.; Domingues, E.M.; Mishra, H.; Ohl, C.D. Mitigating cavitation erosion using biomimetic gas-entrapping microtextured surfaces (GEMS). *Sci. Adv.* **2020**, *6*, eaax6192. [CrossRef]
21. Kang, B.Y.; Kang, S.H. Effect of the number of blades on the performance and cavitation instabilities of a turbo- pump inducer with an identical solidity. *J. Mech. Sci. Technol.* **2015**, *29*, 5251–5256. [CrossRef]
22. Choi, C.H.; Kim, J. Study on the Cavitating Flows in a Turbopump Inducer. *J. Propuls. Power* **2015**, *31*, 537–542. [CrossRef]
23. Liu, Y.; Tan, L. Tip clearance on pressure fluctuation intensity and vortex characteristic of a mixed flow pump as turbine at pump mode. *Renew. Energy* **2018**, *129*, 606–615. [CrossRef]
24. Tan, L.; Yu, Z.; Xu, Y.; Liu, Y.; Cao, S. Role of blade rotational angle on energy performance and pressure fluctuation of a mixed-flow pump. *Proc. Inst. Mech. Eng. Part A J. Power Energy* **2017**, *231*, 227–238.
25. Che, B. *Study on Cavitation Mechanism and Passive Control of Hydrofoil Attachment*; Zhejiang University: Hangzhou, China, 2019.
26. Li, J.Y.; Gao, Z.X.; Wu, H.; Jin, Z.J. Numerical Investigation of Methodologies for Cavitation Suppression Inside Globe Valves. *Appl. Sci.* **2020**, *10*, 5541. [CrossRef]
27. Shin, S.; Hong, J.W.; Nagarathinam, D.; Ahn, B.K.; Park, S.G. Tip Vortex Cavitation and Induced Noise Characteristics of Hydrofoils. *Appl. Sci.* **2021**, *11*, 5906. [CrossRef]
28. Kim, S.J.; Choi, Y.S.; Cho, Y.; Choi, J.W.; Kim, J.H. Numerical Analysis of Inter-Blade Vortex Cavitation Characteristics in a Francis Hydro Turbine. In Proceedings of the 2018 Annual Fall Meeting of the Korean Society for New&Renewable Energy, Jeju, Korea, 4–6 April 2018.
29. Chen, K.; Wan, D. Numerical simulation of hydrofoil cavitation flow based on viscosity modified SST k-ω model. *Hydrodyn. Res. Prog.* **2019**, *34*, 224–231.
30. Yang, D.D.; Yu, A.; Ji, B.; Zhou, J.J.; Luo, X.W. Numerical analyses of ventilated cavitation over a 2-D NACA0015 hydrofoil using two turbulence modeling methods. *J. Hydrodyn.* **2018**, *30*, 101–112. [CrossRef]
31. Wang, X. *Theory and Application of Cavitation Bubble and Supercavitation Bubble Flow*; National Defense Industry Press: Beijin, China, 2009.
32. Upadhyay, M.; Seo, M.W.; Naren, P.R.; Park, J.H.; Nguyen, T.D.B.; Rashid, K.; Lim, H. Experiment and multiphase CFD simulation of gas-solid flow in a CFB reactor at various operating conditions: Assessing the performance of 2D and 3D simulations. *Korean J. Chem. Eng.* **2020**, *37*, 2094–2103. [CrossRef]

applied sciences

MDPI

Article

Numerical Modelling of a Floating Wind Turbine Semi-Submersible Platform

Lander Galera-Calero [1,†], Jesús María Blanco [1,*,†] and Gregorio Iglesias [2,†]

1 Department of Energy Engineering, School of Engineering, University of the Basque Country (UPV/EHU), Plaza Ingeniero Torres Quevedo, Building 1, 48013 Bilbao, Spain; lander.galera@ehu.es
2 Environmental Research Institute, School of Engineering, University College Cork, Civil Engineering Building, 60 College Road, T23XE10 Cork, Ireland; gregorio.iglesias@ucc.ie
* Correspondence: jesusmaria.blanco@ehu.es
† These authors contributed equally to this work.

Abstract: A detailed study is undertaken of the computational modelling of a sub-platform for floating offshore wind using the software Star-CCM+ with the application of the RANS approach. First, a mathematical introduction to the governing equations is carried out. Then, the computational grid is defined, and the grid-independence of the solution is verified. A time-dependent study is performed with the selected time-step. Finally, two examples of 3D decay tests in heave of the sub-platform without and with moorings are presented, accompanied by a damping factor study, with the aim of providing a better understanding of the hydrodynamic damping of the platform. Throughout the process, three degrees of freedom (DoFs) are locked due to the limitations imposed by the use of a symmetry plane; this implementation allowed us to reduce the computational cost of each simulation by 50%. Therefore, three DoFs (heave, surge and pitch) are considered. The coupling study, adding a mooring system in the decay tests and the regular wave tests, shows good agreement between the experimental and computational results. The first half-period of the simulations presents a greater discrepancy due to the fact that the damping of the platform is lower in the computational simulation. However, this does not imply that the hydrodynamic damping is underestimated but may be directly related to the lock of various DoFs associated with the hydrodynamic damping.

Keywords: computational fluid dynamics; decay test; regular wave test; verification and validation; FOWT

Citation: Galera-Calero, L.; Blanco, J.M.; Iglesias, G. Numerical Modelling of a Floating Wind Turbine Semi-Submersible Platform. *Appl. Sci.* **2021**, *11*, 11270. https://doi.org/10.3390/app112311270

Academic Editor: Cesare Biserni

Received: 31 October 2021
Accepted: 25 November 2021
Published: 28 November 2021

1. Introduction

Reducing carbon emissions is one of the main challenges that the world must face. This is reflected in the announcements that several governments have made to go carbon neutral in the next few decades [1]. One of the most popular ways to achieve this is the electrification [2] of all the energy sectors. This creates a unique scenario for renewable technologies, which are one of the most important protagonists of this transition.

Among renewable technologies, wind energy has shown its maturity and continuous evolution, with an increase in the size of turbines [3] and expansion from onshore to offshore. Furthermore, in recent years, this transition has reached deeper areas, paving the way for floating systems to exploit a larger, and better-quality, resource while having fewer social impacts [4]. Despite this expansion, floating wind technology has some challenges that must be addressed to optimize its functioning.

Offshore wind turbines are larger and more powerful than their onshore counterparts [5]. However, floating offshore wind systems, composed of the wind turbine, the floating sub-structure and the mooring lines, have a complex behaviour, as they are affected by environmental, aerodynamic, hydrodynamic, and mooring loads. This complexity is reflected in the turbine working at sub-optimal angles and in constant oscillation and movement. In pursuit of a better performance of these systems, several concepts of floating platforms have been developed, such as spar buoys [6,7], semi-submersible platforms [8] and tension leg platforms (TLP) [9], among others.

To study the design and behaviour of these systems, several approaches exist, which are reflected in different software packages [10]. Numerical simulations are used to design these floating sub-structures and reduce the total cost of the installation of this type of structures due to their complexity. Software packages that combine strip theory and panel methods are well-known in the industry, e.g., Orca Flex [11] and FAST [12], but they present some limitations due to their neglect of viscous effects [13,14].

Another approach, which is gaining popularity in recent years due to the increase in computational power, is to apply high-fidelity numerical models such as Computational Fluid Dynamics (CFD), which combined with six degrees of freedom (6DoFs) present accurate solutions that only require the geometry and the mass properties of the platforms. These numerical models, based on the Reynolds-Averaged Navier–Stokes (RANS) equations, are used to overcome the problem of the viscous effects [15]. Some researchers have compared both approaches [8,13], achieving good agreement between both methods. It is important to highlight that there are more approaches in CFD, such as Direct Numerical Simulations (DNS), which directly apply the Navier–Stokes equations. However, this approach is very computationally expensive, and it is mainly done in academia. Another approach for turbulent modelling is the Large Eddy Simulation (LES) approach, which resolves the large-scale turbulence while modelling the small isotropic turbulent scales. For this article, the RANS approach has been used due to its maturity and its good computational cost in relation to the accuracy of the method, besides its very strong performance coupled with the maturity of the turbulent models developed around this approach. Along these lines, comparisons of different approaches for the simulation of a numerical wave flume have been done to have a better understanding of how different types of turbulence modelling could affect the behaviour of regular waves and the reflection process due to the existence of an extinction system [16].

The first studies that were carried out around CFD simulations were focused on cylindrical floaters [17]. In recent years, more complex simulations have been carried out to better understand the behaviour of these types of structures. For example, numerous articles have been published simulating different decay tests of different structures. Furthermore, another way to study these structures has been to focus on the aerodynamics of the turbine when having a defined motion in the platform [18–20]. Moreover, the study of this type of platforms has been extended in recent years, with several studies focused on decay testing for different degrees of freedom [11,21]. Tran et al. [19,22] studied the unsteady aerodynamics of a turbine by imposing a sine function onto a spar platform. In recent years, various articles have been focusing on the complex behaviour of these structures. Liu et al. [22] studied the effects of the movement of the OC4 DeepCWind platform on the aerodynamics of the wind turbine and vice versa [23], work that continued with the development of an aero-hydro-mooring elastic fully coupled tool by inducing a combined wind and wave condition.

Following their work, Tran and Kim [22] investigated a 5MW wind turbine over an OC4 DeepCWind semi-submersible platform by comparing the results of the dynamic fluid body interaction methodology with those obtained with the open-source software FAST. The aero-hydrodynamic performance of the system was studied under coupled wind-wave conditions and in decay testing [24]. Study of the IDEOL platform in regular wave tests by applying a coupled simulation of multi-body systems with CFD and comparison of the results with experimental data was done by Beyer et al. [25]. An unsteady actuator line model for the aerodynamic simulation coupled with a CFD solver for the hydrodynamic part was studied by Cheng et al. [26].

Similar works have been done for floating structures but for different aims. Bi et al. [27] present a work in which they experimentally study the decay testing of a multi-module aquaculture platform where they analysed the motion of the platform while also analysing, apart from calculating the natural periods, the stiffness with the mooring system. Furthermore, they study the response of the platform when facing regular waves. In relation to the experimental study of mooring systems in this type of structures, Zhao et al. [28] carried out deep research on this topic. Finally, joint research of computational and experimental approaches has been done too around this type of structures, focusing on extreme conditions [29].

However, it is important to find a computational mesh that has a good relation between computational cost and accuracy. The latter one is especially important, since a bad mesh can modify the response of the sub-platform analysed.

This work presents in the following section a summary of the mathematical treatment used in this type of problems. Furthermore, the procedure followed in order to obtain the results is defined, to give a better understanding of this research. Finally, Section 3 presents the outputs of the simulations for platforms both with and without moorings, validated through the data obtained by Saitec Offshore Technologies in their experiments in the LIR/NOTF laboratory through the ARCWIND Project (adaptation and implementation of floating wind energy conversion technology for the Atlantic region).

2. Materials and Methods

2.1. Governing Equations

The behaviour of every incompressible Newtonian flow can be described with the mass and momentum conservation equations, Equations (1) and (2), respectively. These equations are Navier–Stokes equations.

$$\frac{\partial u_i}{\partial x_i} = 0 \tag{1}$$

$$\frac{\partial u_i}{\partial t} + u_j \frac{\partial u_i}{\partial x_j} = -\frac{1}{\rho} + \frac{\partial p}{\partial x_i} + g_i + \frac{1}{\rho}\frac{\partial \tau_{ij}}{\partial x_j} \tag{2}$$

where i and j are the indexes for 2D flows, u_i is the time-averaged component of the velocity vector, ρ is the density of the fluid, t the time, p the pressure, g the gravity acceleration and τ_{ij} the viscous stress tensor. However, transfer of these equations to computational modelling is not extended due to the high computational cost. Therefore, all the instant values obtained in the Navier–Stokes equations are decomposed in mean and fluctuating values with the Reynolds decomposition once we apply Reynolds-Averaged Navier–Stokes (RANS), Equations (3) and (4).

$$\nabla \cdot \boldsymbol{U} = 0 \tag{3}$$

$$\frac{\delta(\rho \boldsymbol{U})}{\delta t} + \nabla(\rho(\boldsymbol{U} - \boldsymbol{U}_g))\boldsymbol{U} = -\nabla p_d - \boldsymbol{g}.x\nabla\rho + \nabla\left(\mu_{eff}\nabla \boldsymbol{U}\right) + (\nabla \boldsymbol{U})\cdot\nabla\mu_{eff} + \boldsymbol{f}_s \tag{4}$$

where $\boldsymbol{U}$ represents the flow velocity, $\boldsymbol{U}_g$ is the velocity of the grid nodes, p_d is the dynamic pressure by substracting the hydrostatic component from the total pressure p, and $\boldsymbol{g}$ is the gravity acceleration, ρ is the mixture density of both flows, μ is the effective dynamic viscosity, ν is the kinematic viscosity and ν_t the eddy kinematic viscosity.

With this decomposition, an introduction of new terms appears, creating an open system of equations. Thus, more equations are needed to close them. In this scenario, different turbulence models have been developed throughout the years to solve this problem. To do that, these models associate the Reynolds stresses with the mean flow variables. In this study, the k-ω SST turbulence model, Equations (5) and (6), developed by Menter is used [30].

$$\frac{\partial k}{\partial t} + \boldsymbol{u}_j \frac{\partial k}{\partial x_j} = \frac{\partial}{\partial x_j}\left[\left(\nu + \sigma\frac{k}{\omega}\right)\frac{\partial k}{\partial x_j}\right]\beta^* \cdot k\omega + \tau_{ij}\frac{\partial \boldsymbol{u}_i}{\partial x_j} \tag{5}$$

$$\frac{\partial \omega}{\partial t} + \boldsymbol{u}_j \frac{\partial \omega}{\partial x_j} = \left[\left(\nu + \sigma\frac{k}{\omega}\right)\frac{\partial \omega}{\partial x_j}\right] - \beta k\omega^2 + \frac{\sigma_d}{\omega}\frac{\partial k}{\partial x_j}\frac{\partial \omega}{\partial x_j} + a\frac{\omega}{k}\tau_{ij}\frac{\partial \boldsymbol{u}_i}{\partial x_j} \tag{6}$$

where k is the turbulence kinetic energy, u is the velocity of the flow, ω is the specific dissipation rate, and ν is the kinematic viscosity, while σ, β and a are constants.

However, in this type of problems, more than one fluid is normally simulated, and another tool is needed for the definition of each fluid and for tracking the free surface. In this study, for this aim, the Volume of Fluid (VOF) presented by Hirt and Nichols [31] is used. It is a free-surface modelling technique, i.e., a numerical technique for tracking and locating the free surface (or fluid–fluid interface). For each phase considered in the model, a variable is introduced as the volume fraction of the phase in the computational cell. Volume fractions represent the space occupied by each phase, and the laws of conservation of mass and momentum are satisfied by each phase individually. In each of the control volumes, the volume fractions of all phases sum up to unity. This model defines the percentage of volume of a certain cell that is occupied by a fluid. This percentage is defined by the following variable:

$$\alpha_i = \frac{V_i}{V} \tag{7}$$

where α_i is the variable, V_i the volume occupied by the specified fluid and V the volume of the cell. In FOWT simulations, fluids are defined by primary and secondary fluids, which have an index of 1 and 0, respectively. So, when a cell is occupied only with the primary fluid, in this case water, its value will be one, and when only air exists, the value will be zero, and finally when the cell has both fluids, a fraction between 0 and 1 will determine the ratio of volume between them. The α-function behaves as another property of the fluid. Because of that, it satisfies the advection equation [32], which states that the material volumes stay constant along the streamlines:

$$\frac{\partial \alpha}{\partial t} + \nabla \cdot (\boldsymbol{u}\alpha) = 0 \tag{8}$$

The position of the free surface corresponds to the value α = 0.5, while the physical properties of each fluid and their volume fraction are given by [33]:

$$\rho = \alpha_i \rho_{water} + (1 - \alpha_i)\rho_{air} \tag{9}$$

$$\mu = \mu_i \rho_{water} + (1 - \alpha_i)\mu_{air} \tag{10}$$

where ρ_i is the specific density of each fluid and μ_i its viscosity. The position of the free surface in the cell is calculated by imposing a perpendicular line to the maximum descent gradient to the cells in its proximity.

2.2. Mesh

2.2.1. Mesh Domain

The computational domain is a semi-cylinder with a diameter of 5 times the hydrodynamic diameter of the platform (D_H) and a height of 10 times the height of the device. These dimensions are sufficiently large to avoid the effect of wave reflection due to the curved walls while maintaining the same depth as the one used in the experimental runs. The volume is differentiated in two main meshes. The background mesh, which embraces the whole volume of study, and the overset mesh, which is focused on the movement of the platform and the behaviour of the fluid alongside it, has variable cell size depending on the area covered.

In the background mesh, five different areas can be distinguished. In four of them, the cell size is defined using the trimmer anisotropic function, which allows us to modify the cell dimensions in all three axes. The area where the cells are not modified is that corresponding to the air, the one that is not numbered in Figure 1, and that is because it is the least important area for this analysis; therefore, it has the largest cells of the volume of study.

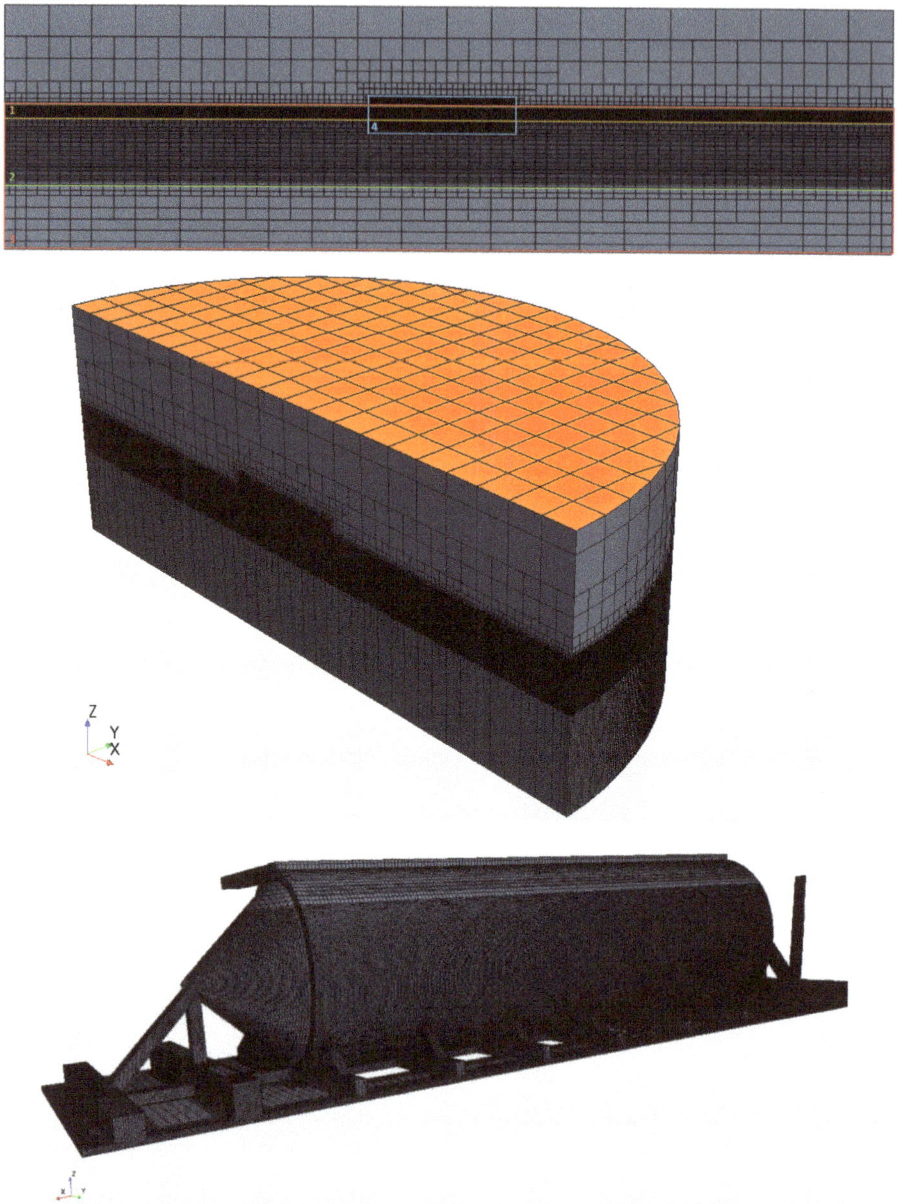

Figure 1. Section with the different sub-volumes used in the background mesh (**top**) and the general volume of study (**mid**) and details of the geometry simulated (**bottom**).

The other 4 areas are defined to have a better understanding of them. Area 1 is the free surface area, which has the smallest cells of the background mesh. The cells' area is defined considering the wavelength, obtained by the heave natural period of the platform, and then the Z dimension is defined by using an aspect ratio (AR) of 4, which is defined in Equation (11). Both horizontal axes, x and y, have the same dimensions in the whole volume.

$$AR = \frac{\Delta z}{\Delta x} \quad (11)$$

Area 2, the under-free-surface refinement space, is refined to not lose energy of the waves in that area formed due to the movement of the platform. The *AR* is maintained in all volumes. The length of its cells in the x and y axes is twice that in the free surface.

The deep-water area, Area 3 in Figure 1, has the same size in the horizontal axes as the air area. The last zone, called the column and defined as Area 4, has the same cell size as the free surface one. This area is created to optimize the data transfer between the overset mesh and the background mesh, and to try to avoid any problems in the mesh definition due to cell size differentiation, which could lead to the breaking of the simulation.

In the background mesh, three types of boundaries can be identified. The top of the volume (the orange face in Figure 1) is defined as a pressure outlet area, which allows the air to flow freely. The curved and bottom boundaries are defined as walls with a non-slip condition, while the vertical boundary in the XZ plane is defined as a symmetry plane, which allows us to only study half of the platform.

This last boundary has an important effect in the reduction of the computational cost of the simulations. However, it only allows us to study three degrees of freedom (DoFs), namely, heave, surge and pitch.

The overset mesh has two different areas: the free surface area, which is designed to cover the free surface interacting with the platform, and the background. Due to this, the free surface area of the overset mesh will be higher than the free surface area of the background mesh. On the contrary, the cell size of it will be much smaller, with its dimensions being 60% smaller than those of the free surface area of the background mesh. The rest of the overset has the same cell size as the cells in Area 4 of the background mesh. Moreover, a prism layer condition is applied around the platform. With a total width of 5 mm and 13 layers with a growth coefficient of 1.2, it ensures that we do not have a value of wall y+ higher than 1.5.

2.2.2. Mesh Independence Study

A mesh convergence study is needed to understand how the mesh definition affects the response of the platform in the different tests, and to achieve the best possible ratio between definition and computational cost. To achieve this aim, 3 grids were designed to compare their results in one of the cases. A decrease in all the dimensions of 50% was presented to know how the response was sensitive to the cell size change. The free surface cells, the smallest ones of the tank mesh, are 50, 100 and 200 times smaller in relation to the height of the device. The overset cells have a height between 50% and 70% of the free surface cell, depending on each of the grids. Thus, 3 grids of clearly different computational weights were obtained, which can be seen in Table 1.

Table 1. Number of total cells for each of the grids considered in the mesh convergence study.

Grid	Number of Cells
Grid 1	2976.088
Grid 2	11,898.568
Grid 3	26,111.669

All the grids have been simulated with the same time-step to focus just on the dependency of the cell size. In Figure 2, it can be seen that there is almost no difference between the 3 grids. However, some tendencies can be seen when analysing the damping ratios of the signals.

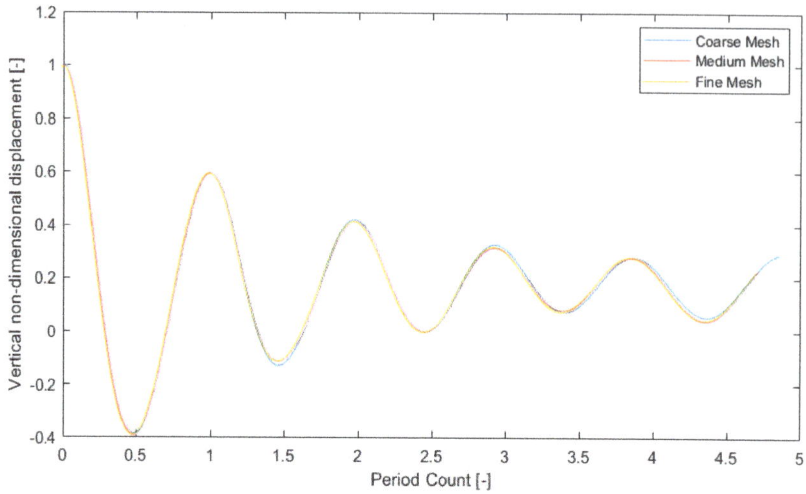

Figure 2. Direct comparison of the signals obtained in the mesh sensitivity study.

The damping ratios can be observed by comparing the mean amplitudes of each of the periods and the decrease in amplitude that can be seen in the period itself. This method is the same as that which Li and Bachinsky-Polic have used [34]. To do that, different equations are needed. The decrease in amplitudes is compared logarithmically:

$$\delta = \log_{10}\left(\frac{\eta_{i+2}}{\eta_i}\right) \tag{12}$$

where η_i is the peak or drought of the decay signal. It is always compared in intervals of 2 to have a comparison between continuous peaks or droughts. The value used as a decrease in the peak or drought value is then used to calculate the damping ratio in that period [31].

$$DR = \sqrt{\frac{1}{1+\frac{\delta}{2\pi}}} \tag{13}$$

Then, to compare the results obtained from this mesh dependency study, the tendency of the points has been defined. A comparison of linear, quadratic and cubic tendencies has been done to see if the highest complexity of the curve has a relationship with the degrees of freedom unlocked.

The damping ratios calculated for each of the meshes in the size dependency study can be seen in Figure 3. In it, the different fittings are set to see how each mesh compare to the other two. The different fittings show that similarities between the two finest meshes increase when the grade of the fitting increases. The most similar fittings are when the grade of the fitting is the same as the degrees of freedom unlock in the simulations.

Besides the mesh sensitivity study, a time-step dependency study was run to see which time-step in relation to the period was needed. The first step of this process was to calculate the Courant number to ensure that all time-steps considered were good enough theoretically. It arises in the numerical analysis of explicit time integration schemes, when these are used for the numerical solution. As a consequence, the time-step must be less than a certain time in many explicit time-marching computer simulations; otherwise the simulation produces incorrect results.

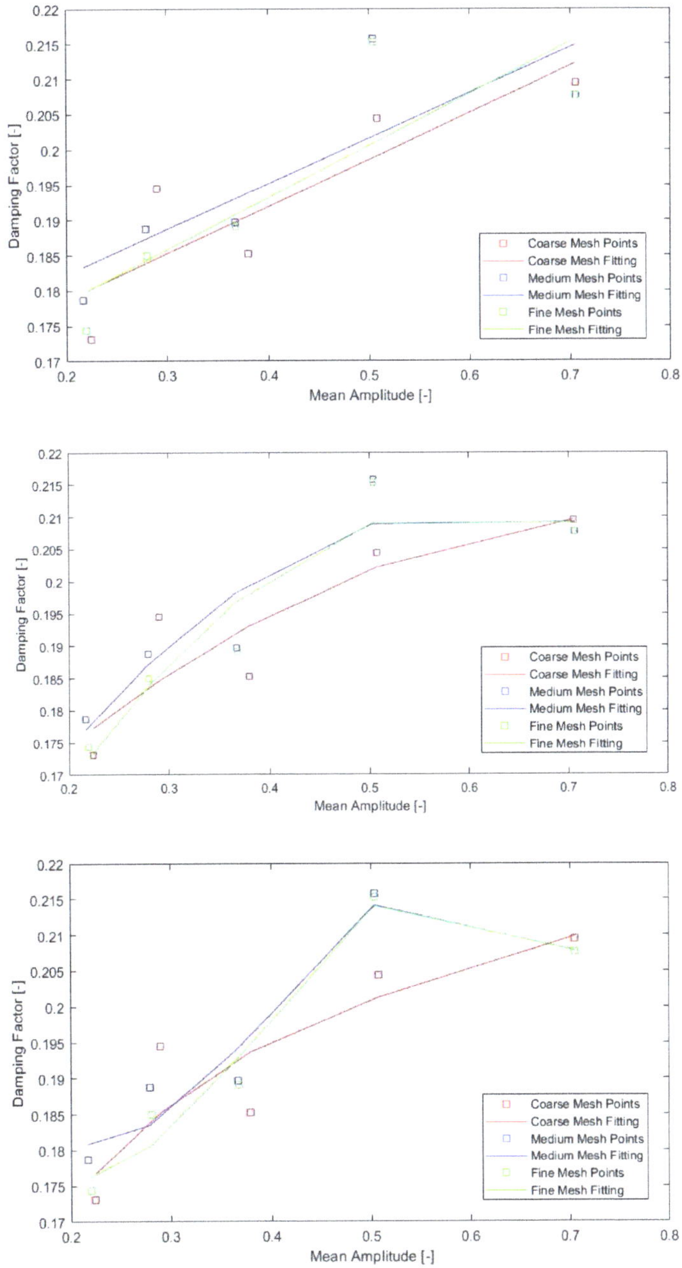

Figure 3. Comparison of the relation of the tendencies used for the case with linear (**up**), quadratic (**mid**) and cubic (**down**) fitting.

This can be seen in Table 2 in which for the three time-steps considered, and the cell size, all the possible Courant numbers in each of the meshes were reasonable. These 3 values are $T_N/1000$, $T_N/1800$ and $T_N/3200$, being T_N the natural period of the degree of freedom analysed. All the values are low, reinforcing the idea of the use of these time-steps, which ensure that the platform will not move more than the height of the smallest cell of the background mesh. In this study, the heave natural period was used. This was done because the natural period of heave is smaller than the other degrees of freedom used in this study.

Table 2. Courant number for every combination of the mesh convergence study.

	T/1000	T/1800	T/3200
Grid 1	0.017	0.0095	0.0052
Grid 2	0.034	0.019	0.0105
Grid 3	0.048	0.0269	0.0148

With this aspect, all the cells, at least, have quality considering the wave celerity and the time-steps of the study. A comparison of the signals obtained in heave motion is presented to see how, with the same starting conditions.

$$CFL = \frac{u\Delta t}{\Delta x} \tag{14}$$

A total of 9 cases were considered in the mesh convergence study. This is the number obtained by combining the 3 grids and 3 time-steps. To see the quality of the general mesh, the Courant number is calculated for each of the meshes. Although there are 9 cases, the time dependency study was carried out in Grid 2, also known as the medium grid.

In Figure 4a small variance in the period of the decay can be seen, although it seems to not affect the amplitudes of the signals.

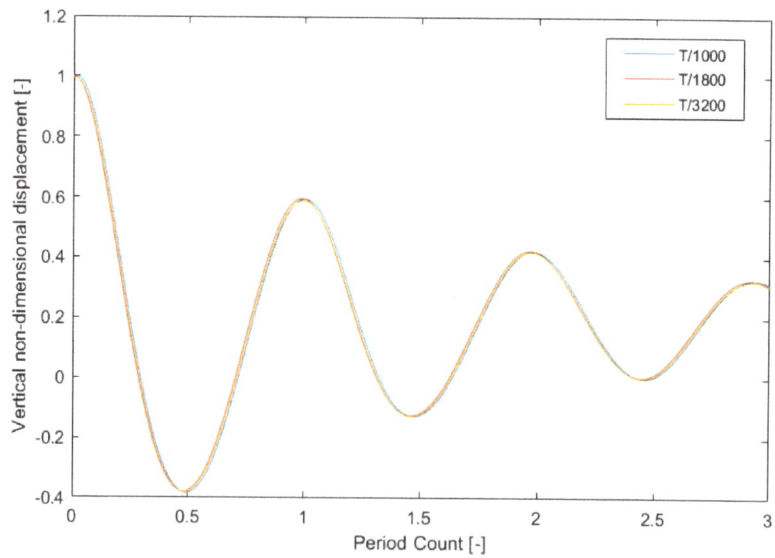

Figure 4. Signal comparison of the different time-steps used in the time dependency study.

Figure 5 shows that the differences in the signals are more significant in the latter periods, which shows that the sensibility of the sub-platform movement is greater at lower velocities.

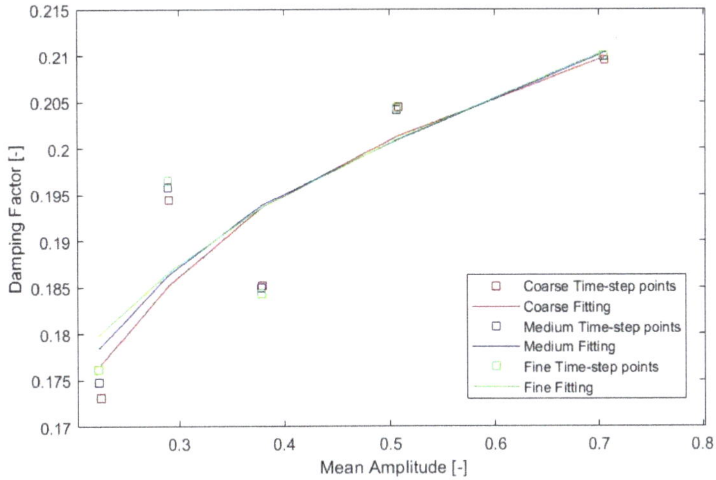

Figure 5. Cubic fitting for the time dependency study where all the time-steps showed similar behaviour with increased differences in the latter periods.

Once the sensitivity study was finished, we decided to use the medium grid and the time-step $T_N/1800$ to obtain the most accurate results possible while keeping the computational cost as low as possible.

3. Results and Discussion

3.1. Decay Tests without Moorings

The first study was to simulate the decay test of the platform in heave without any external forces, such as moorings, that could affect the behaviour of the tests. In the computational runs, only three DoFs are unlocked, surge, heave and pitch, while the rest are locked. This is done due to the symmetry plane imposed in all the simulations to reduce the computational cost of each of the runs.

The first step is to compare the results of the experimental and computational runs over time. To do this, the numerical platform is positioned at the highest point recorded by the experimental runs and then it is allowed to fall, and both signals are compared from their highest point recorded.

In Figure 6, this comparison can be seen clearly. Both signals oscillate around the same equilibrium height, which is a good indication. However, it is clear that the computational run has a lower hydrodynamic damping, as expected. Moreover, the value is similar in the first ones, but starts varying in the latter one. Again, this may be due to the existence of different numbers of DoFs.

However, to have a better understanding of the run, a damping factor study was undertaken. Because of the difference of DoFs in each of the tests, only a linear fitting was done to compare the damping of each of the tests. The slopes of each of the fittings can be related to the damping of each of the experiments. The slope of the experiment is 0.11, while the slope of the numerical fitting is 0.03. Clearly, the damping in the experimental campaign is much higher.

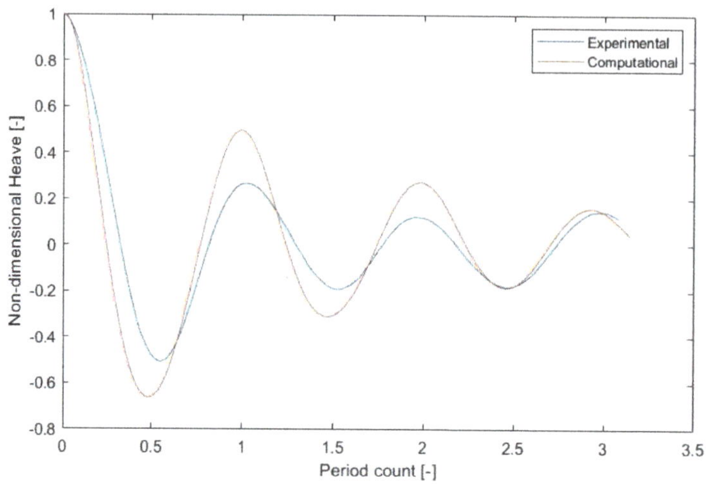

Figure 6. Signal comparison of the experimental (blue) and computational (orange) decays without the interaction of mooring systems.

In Figure 7, it can be seen that the damping factor in the latter period is 0, which means that the amplitude increased in the experimental runs.

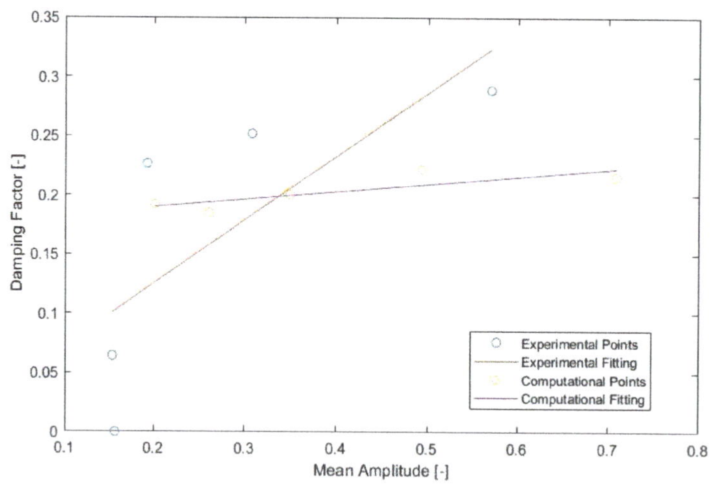

Figure 7. Damping factor study between the experimental and computational tests with linear fitting of both. Larger degree fittings were not studied due to the different DoFs used in both runs.

This is because the amplitudes are very low and the behaviour of the platform in other DoFs that were not modelled in the simulations affects the response in heave. This is very difficult to simulate in the computational domain, and impossible if some DoFs are locked. These results magnify the effect of the damping in the platform, creating a larger slope. Although in the comparison the computational data seem to follow a linear tendency, it is important to remember that a cubic fitting adapts much better to its results, due to the three DoFs used in the runs. However, the different numbers of DoFs between the experimental and computational runs make this fitting less interesting.

Figure 8 shows this by deleting the last two damping factors, related to the last period, which show a clearer relation between the slopes, where the experimental test gives a slope of 0.0467 while the numerical one has a slope of the value 0.0287. The difference in height is, again, given by the lesser damping in the numerical models; however, the relation between periods is more logical. This makes it clear that for further research, a much longer time of decay testing will be needed, apart from the study of different DoFs.

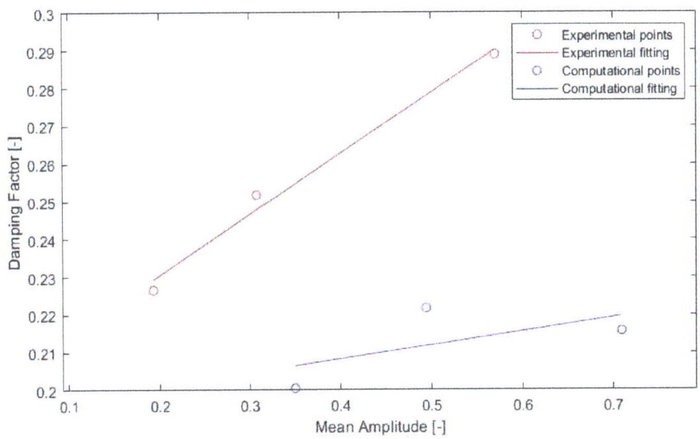

Figure 8. Damping factor comparison with the data from the first periods in order to delete the more significant effect of DoFs that were not unlocked in the computational tests.

3.2. Decay Test with Moorings

To have more data to compare and to test the quality of the meshes, a decay test with mooring lines was performed. This gives us the possibility to, once again, comparing our simulations with the experimental tests and also to testing the same uncertainty with different numbers of DoFs active but with lower impact due to the existence of the mooring lines.

Thanks to the mooring lines behaving symmetrically as springs for this particular set of tests, we were able to keep the symmetry plane in the computational tests. The lines are positioned with a gap of 90 degrees between them and are displaced 45 degrees to the x-axis. Moreover, the lines are parallel to the XY (horizontal) plane.

The springs are connected to the same point of the platform as in the experimental tests and to the curved wall of the volume of study in the simulations, with the sole aim of maintaining the same position in relation to the platform.

Because of the existence of the mooring lines in the experimental tests, which work to maintain the platform in the desired position, a lower effect of the DoFs that are locked in the simulations is observed, showing much better matching between the experimental and computational tests.

Figure 9 shows the comparison over time of the decay test with mooring lines, in which a closer relationship between the two lines can be clearly seen. Moreover, the experimental signal keeps decreasing in the third period, which may be related to a decrease in the influence of the effect of roll in these tests.

The computational signal maintains the same period throughout all the experiments, with a negligible error in the period of less than 1%, a value that can be seen in the decay tests without moorings as were shown previously in Figure 6.

The tendency for larger amplitudes in the computational runs is again reproduced. The damping factor study is done for this type of decays, showing a much better relation than in the runs without moorings.

Figure 10 shows that both fittings are much closer than in the comparisons without moorings. The slope of the experimental test is 0.0155, while that of the computational test is 0.022. This is contrary to the tendency found for the case without moorings, however, which shows less total damping because it stays below the experimental fit.

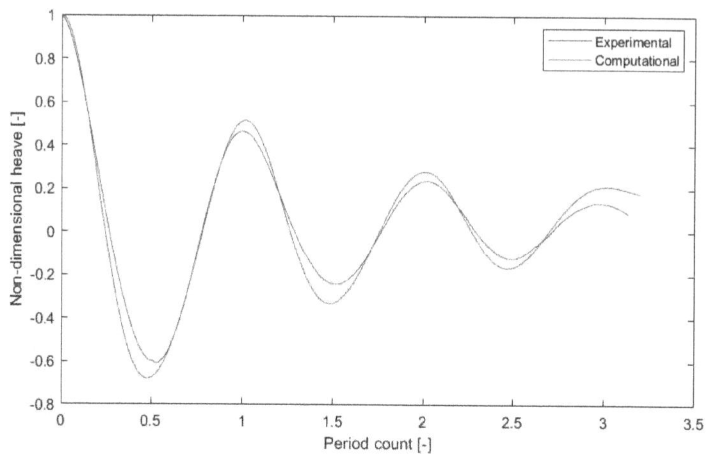

Figure 9. Signal comparison of the experimental (blue) and computational (orange) decays with the interaction of mooring systems.

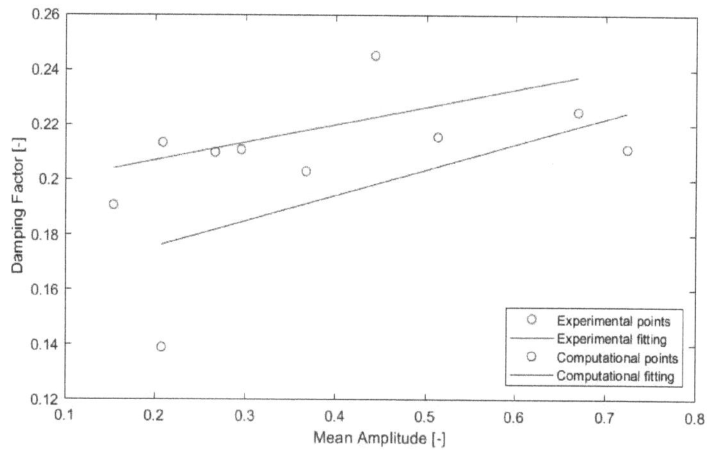

Figure 10. Damping factor study between the experimental and computational tests with linear fitting of both.

4. Conclusions

In this article, we have focused on the process of the first steps to analyse floating platforms for offshore wind, aiming to start a whole computational simulation of the design and verification process for this type of structures. We started with mesh and time dependency studies, and then we analysed a few of the first steps needed for this type of structures, such as decay tests and regular wave tests. To be able to compare the numerical modelling, the motion obtained from experimental tests in the LIR/NOTF wave basin by Saitec Offshore Technologies was used as the ground truth.

The main aim of this study was to consolidate a good basis for both hydrodynamic and computational understanding of this type of simulations before proceeding to more complex steps such as towing tank testing including regular and irregular wave tests.

Throughout the process, three DoFs were locked due to the limitations of the use of a symmetry plane. However, this implementation allowed us to decrease the computational weight of each of the simulations is half. Because of that, only heave, surge and pitch DOFs were used. The coupling study, adding a mooring system in the decay tests and the regular wave tests, showed good relations between the experimental and computational results.

Although the results shown in the study are promising, some clear improvements can be made in future research to close the differences shown in our comparisons.

The damping in the simulations at low velocities, peaks and droughts is lower in the computational runs than in the experimental ones. This is related to the dynamic viscosity, which should vary with the movement of the platform instead of being considered constant. We believe that the difference is more visible in a platform like the studied one with a complex geometry.

The first half-period of the simulations has a greater discrepancy in time due to the damping, because in those first moments the damping of the platform is lower in the computational domain. However, this does not imply that we are underestimating the hydrodynamic damping but may be directly related to the lock of various DoFs, defining the hydrodynamic damping in those DoFs.

Although the application of the symmetry plane is commonly used for this type of research, the lock of three DoFs may be directly related to the hydrodynamic damping of the platform, which allows the researchers some margin of error.

Moreover, the existence of sensors in the experimental prototype imposes a small difference in its behaviour that cannot be simulated.

In future research, more detailed analysis will be conducted on the decay testing, and the mesh and time dependency studies will be extended to regular and irregular waves, as well as to currents and towing tests, which are of great importance for this type of structures. Finally, the simulations show promising results, and the researchers feel positive regarding the implementation of this type of simulations in industrial processes, which can be of great help when studying the behaviour of platforms taking into consideration the non-linearity with greater detail.

Author Contributions: Conceptualization, L.G.-C.; methodology, validation, G.I.; formal analysis, J.M.B. and G.I.; data treatment, L.G.-C.; writing—original draft preparation, L.G.-C.; writing—review and editing, J.M.B. and G.I.; supervision, J.M.B. and G.I. All authors have read and agreed to the published version of the manuscript.

Funding: The current investigation was developed under the framework of the European Regional Development Fund through the "Interreg Atlantic Area Programme" under contract EAPA 344/2016, providing experimental inputs to complete this study.

Institutional Review Board Statement: Not applicable.

Informed Consent Statement: Not applicable.

Data Availability Statement: Not applicable.

Acknowledgments: The authors would like to thank the University of the Basque Country through the research group (GIU19/029)-(IT1314-19) and Saitec Offshore Technologies for their valuable guidance and data provided for the validations through the ARCWIND Project (adaptation and implementation of floating wind energy conversion technology for the Atlantic region).

Conflicts of Interest: The authors declare no conflict of interest.

Abbreviations

AR	Aspect Ratio
CFL	Courant–Friedrichs–Lewis number
DoF	Degree of Freedom
DR	Damping Ratio
FOWT	Floating Offshore Wind Turbine
RANS	Reynolds-Averaged Navier–Stokes
TLP	Tension Leg Platforms
VOF	Volume of Fluid

References

1. Carbon Neutral Government Programme. Available online: https://environment.govt.nz/what-government-is-doing/key-initiatives/carbon-neutral-government-programme/ (accessed on 15 June 2021).
2. Jacobson, M.Z.; Delucchi, M.A.; Bauer, Z.A.F.; Goodman, S.C.; Chapman, W.E.; Cameron, M.A.; Bozonnat, C.; Chobadi, L.; Clonts, H.A.; Enevoldsen, P.; et al. 100% Clean and Renewable Wind, Water, and Sunlight All-Sector Energy Roadmaps for 139 Countries of the World. *Joule* **2017**, *1*, 108–121. [CrossRef]
3. Serrano-González, J.; Lacal-Arántegui, R. Technological Evolution of Onshore Wind Turbines—A Market-Based Analysis. *Wind Energy* **2016**, *19*, 2171–2187. [CrossRef]
4. Rinaldi, G.; Garcia-Teruel, A.; Jeffrey, H.; Thies, P.R.; Johanning, L. Incorporating Stochastic Operation and Maintenance Models into the Techno-Economic Analysis of Floating Offshore Wind Farms. *Appl. Energy* **2021**, *301*, 117420. [CrossRef]
5. Li, H.; Guedes Soares, C.; Huang, H.-Z. Reliability Analysis of a Floating Offshore Wind Turbine Using Bayesian Networks. *Ocean Eng.* **2020**, *217*, 107827. [CrossRef]
6. Tian, X.; Xiao, J.; Liu, H.; Wen, B.; Peng, Z. A Novel Dynamics Analysis Method for Spar-Type Floating Offshore Wind Turbine. *China Ocean Eng.* **2020**, *34*, 99–109. [CrossRef]
7. Nematbakhsh, A.; Olinger, D.J.; Tryggvason, G. Nonlinear Simulation of a Spar Buoy Floating Wind Turbine under Extreme Ocean Conditions. *J. Renew. Sustain. Energy* **2014**, *6*, 033121. [CrossRef]
8. Hu, C.; Sueyoshi, M.; Liu, C.; Liu, Y. Hydrodynamic Analysis of a Semi-Submersible Type Floating Wind Turbine. In Proceedings of the 11th (2014) Pacific/Asia Offshore Mechanics Symposium, PACOMS 2014, Shanghai, China, 12–16 October 2014; International Society of Offshore and Polar Engineers: Mountain View, CA, USA, 2014; pp. 1–6.
9. Nematbakhsh, A.; Bachynski, E.E.; Gao, Z.; Moan, T. Comparison of Wave Load Effects on a TLP Wind Turbine by Using Computational Fluid Dynamics and Potential Flow Theory Approaches. *Appl. Ocean Res.* **2015**, *53*, 142–154. [CrossRef]
10. Dunbar, A.J.; Craven, B.A.; Paterson, E.G. Development and Validation of a Tightly Coupled CFD/6-DOF Solver for Simulating Floating Offshore Wind Turbine Platforms. *Ocean Eng.* **2015**, *110*, 98–105. [CrossRef]
11. Ocrcina OrcaFlex Manual. Available online: https://www.orcina.com/webhelp/OrcaFlex/Default.htm (accessed on 23 October 2021).
12. Jonkman, J.; Buhl, M.L. *FAST User Guide*; National Renewable Energy Laboratory: Golden, CO, USA, 2005.
13. Tran, T.T.; Kim, D.-H. The Coupled Dynamic Response Computation for a Semi-Submersible Platform of Floating Offshore Wind Turbine. *J. Wind Eng. Ind. Aerodyn.* **2015**, *147*, 104–119. [CrossRef]
14. Gueydon, S.; Weller, S. Study of a Floating Foundation for Wind Turbines. *J. Offshore Mech. Arct. Eng.* **2013**, *135*, 031903. [CrossRef]
15. Burmester, S.; Vaz, G.; Gueydon, S.; el Moctar, O. Investigation of a Semi-Submersible Floating Wind Turbine in Surge Decay Using CFD. *Ship Technol. Res.* **2020**, *67*, 2–14. [CrossRef]
16. Galera-Calero, L.; Blanco, J.M.; Izquierdo, U.; Esteban, G.A. Performance Assessment of Three Turbulence Models Validated through an Experimental Wave Flume under Different Scenarios of Wave Generation. *J. Mar. Sci. Eng.* **2020**, *8*, 881. [CrossRef]
17. Quallen, S.; Xing, T.; Carrica, P.; Li, Y.; Xu, J. CFD Simulation of a Floating Offshore Wind Turbine System Using a Quasi-Static Crowfoot Mooring-Line Model. *J. Ocean Wind Energy* **2014**, *1*, 10.
18. Tran, T.T.; Kim, D.-H. A CFD Study into the Influence of Unsteady Aerodynamic Interference on Wind Turbine Surge Motion. *Renew. Energy* **2016**, *90*, 204–228. [CrossRef]
19. Khosravi, M.; Sarkar, P.; Hu, H. An Experimental Investigation on the Performance and the Wake Characteristics of a Wind Turbine Subjected to Surge Motion. In *33rd Wind Energy Symposium*; AIAA SciTech Forum; American Institute of Aeronautics and Astronautics: Reston, VA, USA, 2015.
20. Burmester, S.; Vaz, G.; el Moctar, O. Towards Credible CFD Simulations for Floating Offshore Wind Turbines. *Ocean Eng.* **2020**, *209*, 107237. [CrossRef]
21. Tran, T.; Kim, D.; Song, J. Computational Fluid Dynamic Analysis of a Floating Offshore Wind Turbine Experiencing Platform Pitching Motion. *Energies* **2014**, *7*, 5011–5026. [CrossRef]
22. Liu, Y.; Xiao, Q. Development of a Fully Coupled Aero-Hydro-Mooring-Elastic Tool for Floating Offshore Wind Turbines. *J. Hydrodyn.* **2019**, *31*, 21–33. [CrossRef]
23. Liu, Y.; Xiao, Q.; Incecik, A.; Peyrard, C.; Wan, D. Establishing a Fully Coupled CFD Analysis Tool for Floating Offshore Wind Turbines. *Renew. Energy* **2017**, *112*, 280–301. [CrossRef]

24. Tran, T.T.; Kim, D.-H. Fully Coupled Aero-Hydrodynamic Analysis of a Semi-Submersible FOWT Using a Dynamic Fluid Body Interaction Approach. *Renew. Energy* **2016**, *92*, 244–261. [CrossRef]
25. Beyer, F.; Choisnet, T.; Kretschmer, M.; Cheng, W. Coupled MBS-CFD Simulation of the IDEOL Floating Offshore Wind Turbine Foundation Compared to Wave Tank Model Test Data. In *Proceeding of the Twenty-Fifth International Ocean and Polar Engineering Conference, Kona, HI, USA, 21–26 June 2015*; International Society of Offshore and Polar Engineers: Kona, HI, USA, 2015.
26. Cheng, P.; Huang, Y.; Wan, D. A Numerical Model for Fully Coupled Aero-Hydrodynamic Analysis of Floating Offshore Wind Turbine. *Ocean Eng.* **2019**, *173*, 183–196. [CrossRef]
27. Bi, C.-W.; Ma, C.; Zhao, Y.-P.; Xin, L.-X. Physical Model Experimental Study on the Motion Responses of a Multi-Module Aquaculture Platform. *Ocean Eng.* **2021**, *239*, 109862. [CrossRef]
28. Zhao, Y.; Guan, C.; Bi, C.; Liu, H.; Cui, Y. Experimental Investigations on Hydrodynamic Responses of a Semi-Submersible Offshore Fish Farm in Waves. *J. Mar. Sci. Eng.* **2019**, *7*, 238. [CrossRef]
29. Liu, H.-F.; Bi, C.-W.; Zhao, Y.-P. Experimental and Numerical Study of the Hydrodynamic Characteristics of a Semisubmersible Aquaculture Facility in Waves. *Ocean Eng.* **2020**, *214*, 107714. [CrossRef]
30. Menter, F.R. Zonal Two Equation kw Turbulence Models for Aerodynamic Flows. In Proceedings of the 23rd Fluid Dynamics, Plasmadynamics, and Lasers Conference, Orlando, FL, USA, 6–9 July 1993.
31. Hirt, C.W.; Nichols, B.D. Volume of Fluid (VOF) Method for the Dynamics of Free Boundaries. *J. Comput. Phys.* **1981**, *39*, 201–225. [CrossRef]
32. Katopodes, N.D. Chapter 12—Volume of Fluid Method. In *Free-Surface Flow*; Katopodes, N.D., Ed.; Butterworth-Heinemann: Oxford, UK, 2019; pp. 766–802, ISBN 978-0-12-815485-4.
33. Lu, L.; Tan, L.; Zhou, Z.; Zhao, M.; Ikoma, T. Two-Dimensional Numerical Study of Gap Resonance Coupling with Motions of Floating Body Moored Close to a Bottom-Mounted Wall. *Phys. Fluids* **2020**, *32*, 092101. [CrossRef]
34. Li, H.; Bachynski-Polić, E.E. Validation and Application of Nonlinear Hydrodynamics from CFD in an Engineering Model of a Semi-Submersible Floating Wind Turbine. *Mar. Struct.* **2021**, *79*, 103054. [CrossRef]

applied sciences

MDPI

Article

Determination of the Reference Temperature for a Convective Heat Transfer Coefficient in a Heated Tube Bank

Stanislav Kotšmíd * and **Zuzana Brodnianská**

Faculty of Technology, Technical University in Zvolen, Studentska 26, 960 01 Zvolen, Slovakia; zuzana.brodnianska@tuzvo.sk
* Correspondence: stanislav.kotsmid@tuzvo.sk

Abstract: The paper presents a theoretical analysis of heat transfer in a heated tube bank, based on the Nusselt number computation as one of the basic dimensionless criteria. To compute the Nusselt number based on the heat transfer coefficient, the reference temperature must be determined. Despite the value significance, the quantity has several different formulations, which leads to discrepancies in results. This paper investigates the heat transfer of the inline and staggered tube banks, made up of 20 rows, at a constant tube diameter and longitudinal and transverse pitch. Both laminar and turbulent flows up to Re = 10,000 are considered, and the effect of gravity is included as well. Several locations for the reference temperature are taken into consideration on the basis of the heretofore published research, and the results in terms of the overall Nusselt number are compared with those obtained by the experimental correlations. This paper provides the most suitable variant for a unique reference temperature, in terms of a constant value for all tube angles, and the Reynolds number ranges of 100–1000 and 1000–10,000 which are in good agreement with the most frequently used correlating equations.

Keywords: heat transfer; heat transfer coefficient; tube bank; forced convection; Nusselt number

Citation: Kotšmíd, S.; Brodnianská, Z. Determination of the Reference Temperature for a Convective Heat Transfer Coefficient in a Heated Tube Bank. *Appl. Sci.* **2021**, *11*, 10564. https://doi.org/10.3390/app112210564

Academic Editor: Jesús María Blanco

Received: 9 October 2021
Accepted: 7 November 2021
Published: 10 November 2021

1. Introduction

Heat exchangers consisting of a heated tube bank are widely used in practice to transfer heat from one medium to another. Typically, one fluid flows over the tube bank, while another fluid of a different temperature passes through the tubes. The tubes are arranged as a bank in an inline or staggered manner to heat the surroundings, dissipate excessive heat, or recover waste heat. Tube banks in crossflow are used in many fields, such as industry, biomedicine, mechanisms for heat transfer increase, and nanofluid applications. They can be found in the economizer and evaporator of an air conditioning system [1]. Moreover, they are parts of processes such as biological systems, filtration flow, fibrous media encountered in polymer processing, and in insulation materials [2].

Amatachaya and Krittacom [3] discussed the inline tube bank heat exchanger installed on a gas porous burner, using the air as a working fluid and LPG as a fuel. Gu and Min [4] investigated the thermal-hydraulic characteristics of the bare tube bank and plain finned tube heat exchangers used to cool the compressor bleed air in an aero turbine engine. Wang et al. [5] performed an analysis and optimization of metal-foam tube banks for waste heat recovery from the engine's exhaust gas. Kang et al. [6] investigated the heat transfer and pressure drop of sodium-to-air heat exchanger tube banks on a sodium-cooled fast reactor. A steam reactor powered by a propane fuel consisting of a shell and tube heat exchanger was investigated by Barnoon et al. [7]. Some authors use the nanofluid to improve the heat transfer parameters [8–10].

To evaluate the heat transfer in a tube bank, the Nusselt number is used as one of the basic criteria. It is defined as a ratio of convective to conductive heat transfer across a boundary. In other words, the Nusselt number is a dimensionless heat transfer coefficient. The purpose of using dimensionless numbers is to compare the results of the used quantities independently. There are two well-known ways to obtain the Nusselt number. The first one is based on the heat transfer coefficient, characteristic length, and thermal conductivity; the other one uses the empirical correlations based on the Reynolds and Prandtl numbers. The latter approach is usually mentioned in literature dealing with experimental results. Some authors have created more difficult and complex formulas where other variables are used as well. They are related to the manner in which fluid properties are evaluated. The base is either the free-stream temperature, the film temperature, which is the average of the free-stream and surface temperatures, or the base is the average of the inlet and outlet temperatures, which is not easy to determine without a calculation or experiment.

Due to the lack of computational technology, the first studies of heat transfer were conducted only experimentally and reported the tube spacing effect for banks in crossflow. Grimison [11] and Pierson [12] created the correlations for various tube bank arrangements in the Reynolds number range of 2×10^3 and 4×10^4, while Huge [13] tested different tube diameters at Reynolds numbers ranging from 2×10^3 to 7×10^4. He confirmed the similarity principle applied to the tube banks, despite some departure from the true geometric similarity in the ratio of length to the diameter or to the intertube space. Brevoort and Tifford [14] detailed the flow conditions in a bank of 20 staggered circular tubes with a Reynolds numbers of over 2×10^4. McAdams [15] discussed the effect of the Prandtl number and large differences in temperature-dependent properties in a wide range of the Reynolds numbers from laminar to turbulent flow. The most extensive and precise measurement of the local heat transfer coefficient around a cylinder was carried out by Schmidt and Werner [16]. Kays and Lo [17] extended the available data for a normal gas flow to the small tube diameter banks of various staggered arrangements to lower the Reynolds numbers from 10^3 to 10^4. In the paper by Welch and Fairchild [18], heat transfer coefficients are obtained for the individual rows of the ten-row inline tube banks, under various pitch ratio arrangements.

To study heat transfer rates in the crossflow over the smooth tube banks, the correlating equations obtained by Žukauskas and his co-authors are widely used. One of these investigations, by Samoshka et al. [19], was of a closely spaced staggered tube bank of large smooth tubes in water streams within 21 turbulent regions. They found out that the efficiency of the banks from an energetic point of view increased as the tube spacing decreased. Žukauskas [20] reported the pressure and hydrodynamic resistance of single tubes and tube banks of various arrangements in flows of gases and v (viscous liquids) at higher Reynolds numbers and various Prandtl numbers from 0.7 to 500. The effect of the fluid properties was considered by Žukauskas and Ulinskas [21,22] for the inline and staggered tube banks in a water crossflow at Reynolds numbers ranging from 5×10^4 to 2×10^6 and Prandtl numbers from 3 to 7. They determined the optimal arrangements and geometries of the tube banks in an oil crossflow at Reynolds numbers ranging from 1 to 2×10^4. The former Grimison correlations were modified by Hausen [23], where the empirical formulas were created instead of a graphical representation.

Ramezanpour et al. [24] used the Ansys Fluent software and the RNG k-ε turbulence model with a modified dissipation term in the ε equation, where the Reynolds numbers (based on the maximum mean velocity inside a tube bank and the tube hydraulic diameter) of 10^3, 5×10^3, 10^4, and 10^5 were used. According to the results, the optimal tube spacing for a staggered tube array in crossflow was found for S_T/D = 1.5, 1.3 and S_L/D = 1.15, 1.05, respectively (Figures 1 and 2). Khan et al. [25,26] presented the models for the inline and staggered arrangement of tube banks applicable over a wide range of Reynolds and Prandtl numbers, as well as the longitudinal and transverse pitch ratios. It was determined that the average heat transfer coefficient for tube banks in crossflow depends on the number of longitudinal rows, longitudinal and transverse pitch ratios, and the Reynolds and Prandtl numbers. Moreover, it was proved that the staggered arrangement gave higher heat transfer rates than the inline one. The same findings were observed by Haider et al. [27].

Kim [28] indicated that the heat transfer coefficient might be reduced by 37.1% from the prediction of a well-known correlation by Žukauskas, as the longitudinal pitch decreased.

The heat transfer degradation can be estimated by using an empirical correction factor, and the Žukauskas correlation can predict existing experimental data when the correlation is combined with the empirical factor developed in the study. Yilmaz et al. [29,30] extended and developed the formulas of Gnielinski, Gaddis, and Žukauskas for the calculation of the Nusselt number in tube banks. The main contribution is a unique formula for a wide range of the Reynolds numbers, while the Žukauskas correlations are divided into the Reynolds number subintervals. Niemelä et al. [31] created the CFD model by comparing different boundary conditions, domain dimensions, and turbulence models for unsteady simulations of the inline tube banks. The correlation of Gnielinski is recommended for the tube banks with large transverse spacing as it agrees, within ±13%, with the numerically obtained values.

The former way to compute the Nusselt number based on the heat transfer coefficient is derived from the Fourier and Newton law. As a part of the formula, the reference temperature should be expressed. Depending on the solved problem, the reference temperature is taken at different locations. In a heat exchanger tube, it could be defined as an average bulk fluid temperature at the axial position, according to Córcoles et al. [32]. In some cases, it is taken far from the surface and defined as the inlet temperature. In other cases, it could be defined as a bulk temperature or film temperature. According to Beale [33], some authors, such as Žukauskas, used the temperature in a free stream at the beginning of the subdomain. In the same way, the bulk temperature of the fluid entering the tube is used in the paper by Ge et al. [34], who optimized the shape of the staggered tube bank. Alternatively, authors such as Gnielinski [35] or Bergelin [36] used the log-mean temperature difference, while Le Feuvre [37] and Massey [38] use the midway subdomain temperature. Furthermore, the mass weighted average temperature of the inlet and outlet, with subsequent averaging, was used in the investigation by Castro et al. [39], while Mangrulkar et al. [40] used the temperature of the fluid past the corresponding cylinder. Muzaffar et al. [41] used the logarithmic mean temperature difference when studying the heat transfer of a half-cycle air condition system, using liquefied petroleum gas. The same approach was used to calculate the average air-side heat transfer coefficient for a tube bundle heat exchanger with a novel fin design in the paper by Unger et al. [42]. Xu et al. [43] used a linear interpolation of the average temperature at the inlet and outlet of the channel when evaluating the mean Nusselt number of a staggered array of Kagome lattice structures. Wang et al. [44] proved that an incorrect reference temperature of the fin side surface heat transfer coefficient leads to discrepancies between the experimental and numerical results.

The manuscript points out a need for the correct determination of the reference temperature to calculate the heat transfer coefficient and Nusselt number in a heated tube bank. When considering high Reynolds numbers, it is stated that either of the former methods converge to the unique value. The problem appears at low Reynolds numbers, where no single reference temperature is assumed. In this paper, some of the mentioned locations will be tested to determine the reference temperature, and the resulting Nusselt number will be compared with those obtained by the correlating equations. As a part of the reverse investigation, several representative temperatures for a subdomain were created and compared. Finally, the most suitable reference value with the lowest discrepancy to the experimental measurements is provided in terms of a constant for all tube angles and the Reynolds number ranges of 100–1000 and 1000–10,000.

2. Numerical Investigation

The numerical analyses were performed on the inline and staggered tube banks, which are made up of 20 rows with tube diameters of D = 20 mm, and a longitudinal and transverse pitch $S_L = S_T$ = 40 mm. The wall and inlet temperatures were constantly set to T_w = 333.15 K and T_∞ = 293.15 K, respectively, while the inlet velocities U_∞ corresponded to the Reynolds number in the range of 10^2–10^4. Due to a symmetrical character of the task, the domains shown in Figures 1 and 2 are considered.

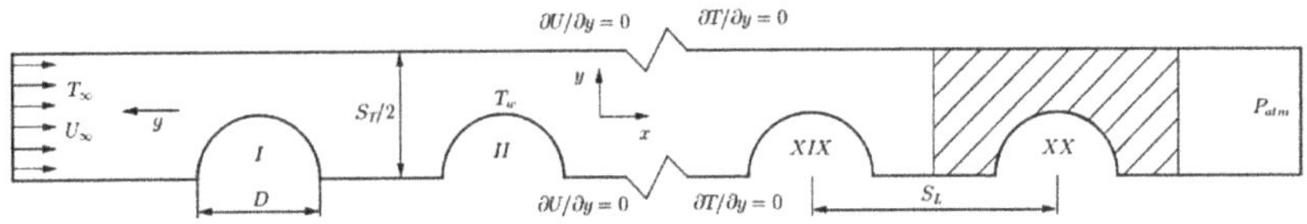

Figure 1. The inline tube bank computational domain.

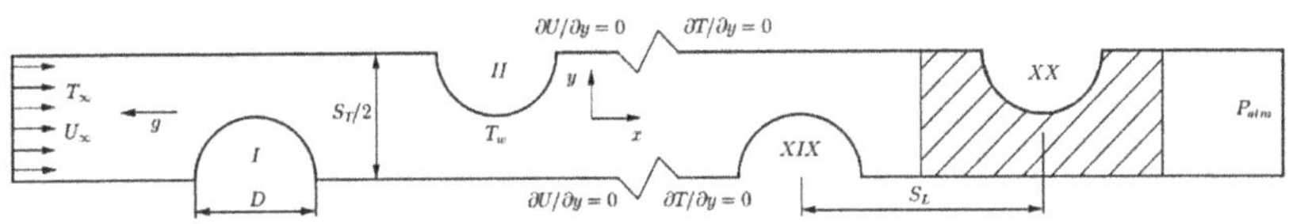

Figure 2. The staggered tube bank computational domain.

The flow is considered to be two-dimensional, steady, and incompressible. The governing equations of mass, momentum, and energy conservation are expressed in terms of the velocity vector v, density ρ, static pressure p, gravity g, dynamic viscosity μ, specific heat capacity c_p, temperature T, and thermal conductivity of fluid k_f as follows:

$$\nabla\cdot\vec{v} = 0 \tag{1}$$

$$\rho\vec{v}\cdot\nabla\vec{v} = -\nabla p + \rho g + \mu\Delta\vec{v} \tag{2}$$

$$\rho c_p\vec{v}\cdot\nabla T = k_f\Delta T \tag{3}$$

The boundary conditions in terms of the velocities U, V (x and y direction) are applied:

$$\text{inlet: } U = U_\infty, \quad V = 0, \quad T = T_\infty \tag{4}$$

$$\text{outlet}: \ P = P_{atm}, \quad T = T_\infty \tag{5}$$

$$\text{tubes}: \ U = 0, \quad V = 0, \quad T = T_w \tag{6}$$

$$\text{symmetry}: \ \frac{\partial U}{\partial y} = 0, \quad V = 0, \quad \frac{\partial T}{\partial y} = 0 \tag{7}$$

Governing Equations (1)–(3) considering the boundary conditions (4)–(7) were solved by the finite volume method using the Ansys Fluent software. The steady pressure-based analyses with the variable air properties and incompressible ideal gas density model were performed. Regarding the variable properties, specific heat c_p (Jkg^{-1}K^{-1}), thermal conductivity k_f (Wm^{-1}K^{-1}), and dynamic viscosity μ (kgm^{-1}s^{-1}) are considered as polynomial functions of temperature according to the standard [45]:

$$c_p(T) = 3.34\cdot10^{-4}T^2 - 0.156T + 1023.53 \tag{8}$$

$$k_f(T) = -2.48\cdot10^{-8}T^2 + 8.92\cdot10^{-5}T + 1.12\cdot10^{-3} \tag{9}$$

$$\mu(T) = -3.76\cdot10^{-11}T^2 + 6.95\cdot10^{-8}T + 1.12\cdot10^{-6} \tag{10}$$

To perform the numerical analyses, the Direct Numerical Simulation solver was used for the laminar flow, while the turbulent one was solved by the k-ω SST model, using a low inlet turbulent intensity. The pressure-velocity coupling was handled by the coupled scheme with the pseudo transient formulation. As a spatial discretization, the QUICK schemes were used. The solution was considered to be fully converged when the residuals of continuity, x-velocity, y-velocity, energy, k, and ω parameters met the convergence criterion 10^{-6}. A preliminary study of the mesh grid size was carried out where quadrilateral elements were used. As it is shown in Table 1, five different numbers of elements were created in the wall-adjacent area, and the average Nusselt number for the tube was evaluated. Each mesh variant has a different number and bias of elements in a normal direction to the tube wall in a distance of 1.3 mm, where a boundary layer is predicted. On the basis of this study, the last mesh variant with 14 perpendicular elements was chosen from the accuracy point of view. Furthermore, the y^+ values are included in this study, with the worst values are shown in Table 1. The maximum element height out of the adjacent area is 0.3 mm. Considering the element quality (aspect ratio, angles, etc.), the element widths were appropriately adjusted (Figure 3).

Table 1. Mesh independence test for the tube-adjacent area.

Elements in Adjacent Area	y^+	Average Nu	Error (%)
960	6.153	70.012	-
1280	3.608	57.382	18.04
1600	2.178	56.582	1.39
1920	2.107	56.245	0.60
2240	1.396	56.029	0.39

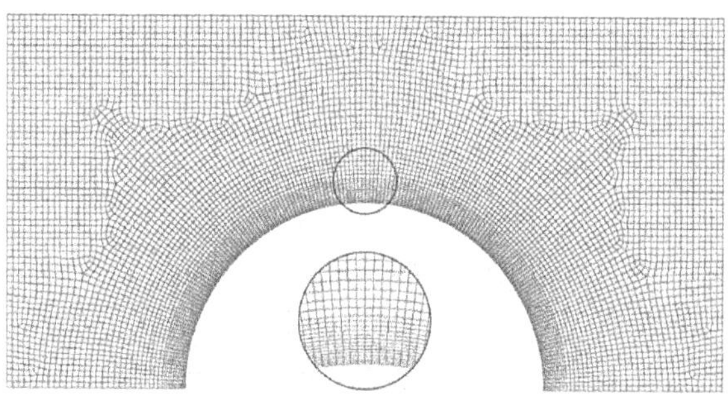

Figure 3. The segment grid.

On the basis of the Fourier and Newton law, the local heat transfer coefficient is defined as:

$$h_\theta = -k_w \frac{dT}{dr}\bigg|_{r=0} \frac{1}{T_w - T_{ref}} \tag{11}$$

The mean or average heat transfer coefficient over the investigated domain is given as:

$$h_{avg} = \frac{1}{A}\int h_\theta dA \tag{12}$$

The Nusselt number over the investigated domain is defined as [46]:

$$Nu = \frac{h_{avg} D}{k_f} \tag{13}$$

where k_w is the thermal conductivity at the tube wall and A is the tube area.

Performing our search for the reference temperature location, several variants were tested according to Figure 4, as described in Table 2. Except the mentioned variants, some combination trials were built, and the most suitable ones are described later.

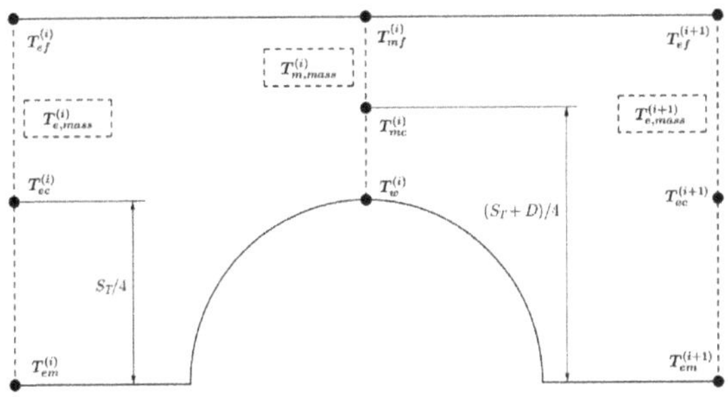

Figure 4. Temperature location layout.

Table 2. The reference temperature characteristics.

Denotation	Description
$T_{ef}^{(i)}$	Temperature for *i*th row in the free stream of the subdomain inlet.
$T_{ec}^{(i)}$	Temperature for *i*th row in the half of the subdomain inlet.
$T_{em}^{(i)}$	Temperature for *i*th row on the bound of the subdomain inlet.
$T_{mf}^{(i)}$	Temperature for *i*th row in the free stream of the subdomain centre.
$T_{mc}^{(i)}$	Temperature for *i*th row in the half of the subdomain centre.
$T_{w}^{(i)}$	Temperature for *i*th row at the tube wall.
$T_{e,mass}^{(i)}$	Mass weighted average temperature for *i*th row at the subdomain inlet.
$T_{m,mass}^{(i)}$	Mass weighted average temperature for *i*th row at the subdomain centre.
$\Delta T_{ef}^{(i)}$	LMTD for *i*th row based on T_{ef}
$\Delta T_{ec}^{(i)}$	LMTD for *i*th row based on T_{ec}
$\Delta T_{em}^{(i)}$	LMTD for *i*th row based on T_{em}
$\Delta T_{e,mass}^{(i)}$	LMTD for *i*th row based on $T_{e,mass}$

The logarithmic mean temperature difference (LMTD) is computed based on the subdomain inlet temperature T_{in}, outlet temperature T_{out}, and wall temperature T_w as follows:

$$\Delta T = \frac{(T_{in} - T_w) - (T_{out} - T_w)}{\ln\left(\frac{T_{in} - T_w}{T_{out} - T_w}\right)} \tag{14}$$

To provide the reference temperature with a better agreement compared to the other variants, the mean of the selected temperature points on the subdomain bound was considered where the formula can be expressed as follows:

$$T_{ref}^{(i)} = \frac{T_{em}^{(i)} + T_{em}^{(i+1)} + T_{mf}^{(i)} + T_{w}^{(i)}}{4} \tag{15}$$

When considering the correlating equations, the Reynolds number is defined with the velocity in the minimum cross-section, while the Prandtl number is defined based on the thermal conductivity (k) location:

$$Re_{max} = \frac{\rho U_{max} D}{\mu} \tag{16}$$

$$Pr = \frac{c_p \mu}{k} \quad (17)$$

The following correlating equations will be compared with the computations:

$$\text{Grimison}: \ Nu = 1.13 C_1 Re_{max}^m Pr^{1/3} \quad (18)$$

$$\text{Zukauskas}: \ Nu = C_2 Re_{max}^m Pr^{0.36} \left(\frac{Pr}{Pr_w}\right)^{1/4} \quad (19)$$

$$\text{Yilmaz}: \ Nu = C_2 Re_{max}^{0.4} \left[1 + \left(\frac{Re_{max}}{p}\right)^m\right]^n Pr^{0.36} \quad (20)$$

$$\text{Hausen}: \ Nu = C_1 C_2 Re_{max}^m Pr^{0.31} \quad (21)$$

$$\text{Khan}: \ Nu = C_2 \left(C_1 Re_{max}^{1/2} Pr^{1/3} + 0.001 Re_{max}\right) \quad (22)$$

$$\text{ESDU}: \ Nu = C_2 Re_{max}^m Pr^{0.34} \left(\frac{Pr_w}{Pr}\right)^{0.26} \quad (23)$$

The arrangement factor C_1 for $S_T/D = S_L/D = 2$ and the correction factors C_2 and m, which are dependent on Re, are noted in Tables 3 and 4.

Table 3. The correlation factors for the inline arrangement.

Author	C_1	C_2	m	n	p
Grimison [11]	0.229	-	0.632	-	-
Žukauskas [20–22]	-	0.52 * 0.27 ×	0.5 * 0.63 ×	-	-
Yilmaz [29,30]	-	0.9	5	0.5	290
Hausen [23]	$1 - 0.0605\sqrt{\frac{1000}{Re_{max}}}$	0.34	0.61	-	-
Khan [25,26]	0.542	1.43	-	-	-
ESDU [47]	-	0.742 △ 0.211 ◊	0.431 △ 0.651 ◊	-	-

* for $Re = 10^2$–10^3; × for $Re = 10^3$–10^4; △ for $Re = 10$–$3{\cdot}10^2$; ◊ for $Re = 3{\cdot}10^2$–$2{\cdot}10^5$.

Table 4. The correlation factors for the staggered arrangement.

Author	C_1	C_2	m	n	p
Grimison [11]	0.482	-	0.556	-	-
Žukauskas [20–22]	-	1.04 * 0.71 × 0.35 !	0.4 * 0.5 × 0.6 !	-	-
Yilmaz [29,30]	-	1.04	1.84	0.125	500
Hausen [23]	1.18	0.35	0.57	-	-
Khan [25,26]	0.567	1.61	-	-	-
ESDU [47]	-	1.309 △ 0.273 ◊	0.36 △ 0.635 ◊	-	-

* for $Re = 0$–$5{\cdot}10^2$; × for $Re = 5{\cdot}10^2$–10^3; ! for $Re = 10^3$–$2{\cdot}10^5$; △ for $Re = 10$–$3{\cdot}10^2$; ◊ for $Re = 3{\cdot}10^2$–$2{\cdot}10^5$.

3. Results and Discussion

A dependence of the Nusselt number on the Reynolds number for the inline tube bank is shown in Figure 5, where the mentioned variants of the reference temperatures were used to compare the computed Nusselt numbers with the correlating equations of Žukauskas, Grimison, Yilmaz, Hausen, Khan, and ESDU standards. Except the variants T_{em}, T_{ec}, ΔT_{em}, and ΔT_{ec}, the other methods, shown hatched, underestimate the experimental results in the investigated area. Although the former methods meet the value order, the course slope is not in agreement with the correlating equations for a significant part of the Re range.

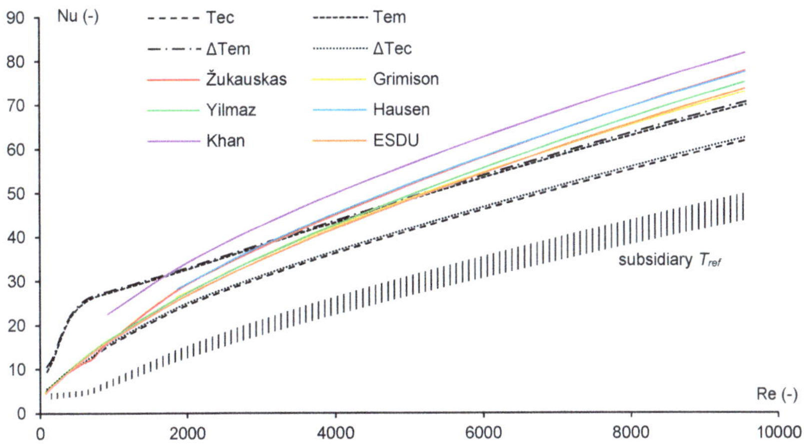

Figure 5. Average Nusselt number for the inline tube bank.

When considering the variant with the lowest discrepancies to any correlating equation, the temperature T_{ec} provides a suitable reference temperature for the Nusselt numbers that are in a good agreement with the Žukauskas correlating equation in the range of $Re = 10^2$–10^3 (Figure 6). The maximum discrepancies appear on the boundary Reynolds numbers (4.5%) due to the Žukauskas discontinuous correlations. The average and maximum discrepancies, within the correlation interval, are 0.8% and 1.9%, respectively.

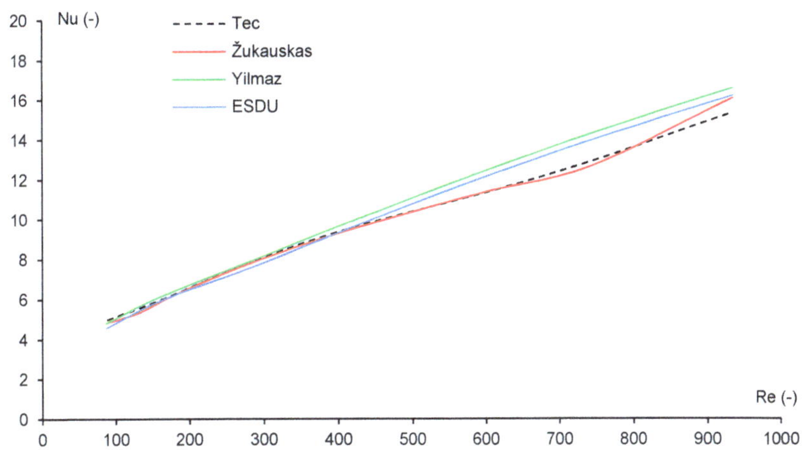

Figure 6. Average Nusselt number for the inline tube bank at Re = 100–1000.

For the turbulent flow over $Re = 10^3$, the approach according to the Equation (15) was tested. As is shown in Figure 7, the results are in good agreement with the Žukauskas correlating equation, where the average and maximum discrepancies are 1.25% and 2.86%, respectively.

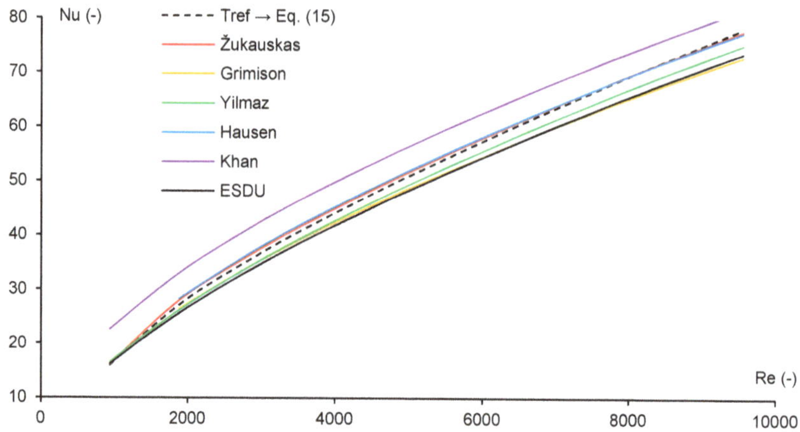

Figure 7. Average Nusselt number for the inline tube bank at Re = 1000–10,000.

In the range of $0 < Re < 10^2$, any constant reference temperature was found to meet the value order and course slope of the mentioned correlating equations, which confirms the hypothesis of the listed authors: that there is no reference temperature which is both representative and easy to compute or measure at low Reynolds numbers.

A dependence of the Nusselt number on the Reynolds number for the staggered tube bank is shown in Figure 8, in the same way as for the inline tube bank. Almost all of the investigated reference temperature variants meet the error range and curve slope of any of the correlating equations. From the accuracy point of view, the temperature T_{mf} is the most suitable to use where the maximum discrepancy for $10^3 < Re < 10^4$ is 2.24% compared to the Žukauskas correlating equation. Due to its range discontinuity, there is no smooth course and therefore the correlation according to the ESDU is taken into consideration for $10^2 < Re < 10^3$. The lowest discrepancies for this variant are obtained when using the temperature $T_{e,\,mass}$, where the highest one is 6.23%. A detailed comparison is shown in Figure 9.

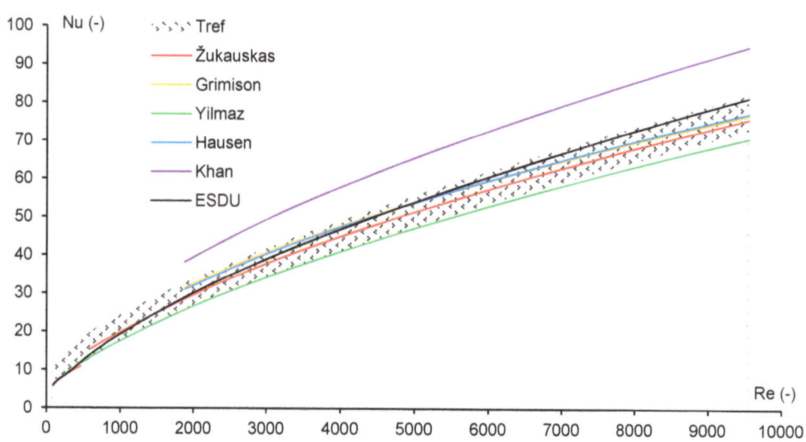

Figure 8. Average Nusselt number for the staggered tube bank.

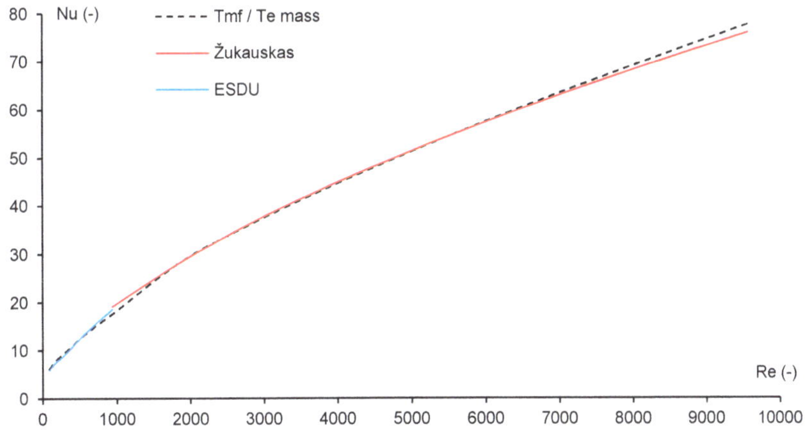

Figure 9. Average Nusselt number for the staggered tube bank—detailed fitting.

The temperature fields (Figure 10) show a thicker boundary layer for lower Reynolds numbers, which causes the diminishing of the temperature gradient on the surface and the reduction in the local and average Nusselt numbers. From the reference temperature point of view, the flow at the staggered tube bank is mixed even for higher Reynolds numbers; therefore, it is simpler to find a unique reference temperature. On the other hand, the inline tube bank has a subdomain divided into the free stream and stream between the tubes, and therefore, there is a need to compile the reference temperatures from several locations for higher Reynolds numbers.

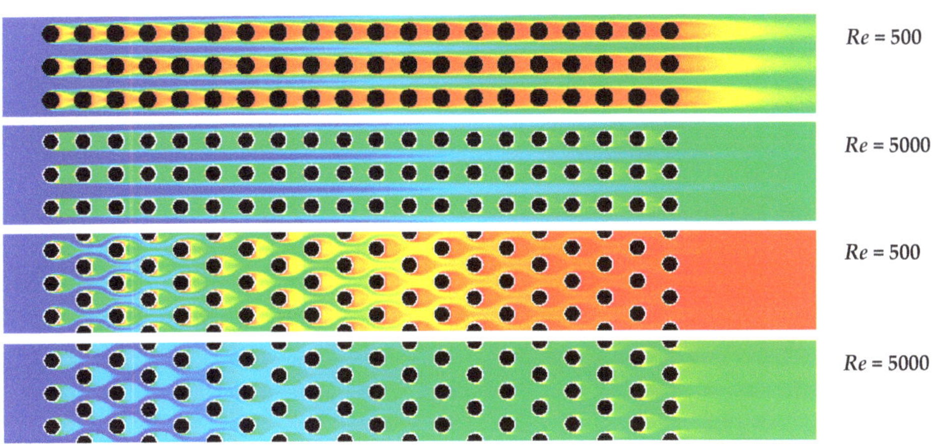

Figure 10. The temperature fields of the inline and staggered tube banks.

The local heat transfer coefficients for the inline and staggered tube banks at $Re = 500$ and 5000 are shown in Figures 11 and 12. As investigated by Žukauskas, the maximum value of the heat transfer for the inline tube bank was observed at $\varphi = 50°$, which is closed to the obtained numerical results. Considering the staggered tube bank, the heat transfer in the first row is similar to that of a single tube. In subsequent rows, an increase in heat transfer is achieved. On the other hand, the heat transfer at low Reynolds numbers is similar for all the tubes. For both the inline and staggered tube banks, the heat transfer becomes similar from the third or fourth row.

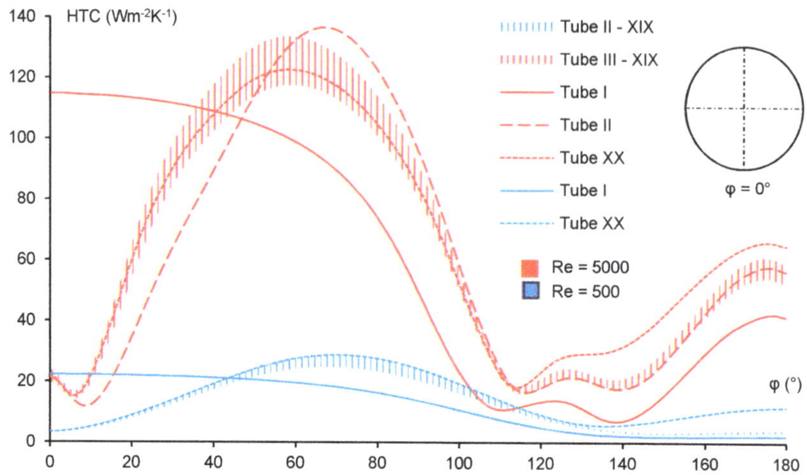

Figure 11. The local heat transfer coefficients of the inline tube bank.

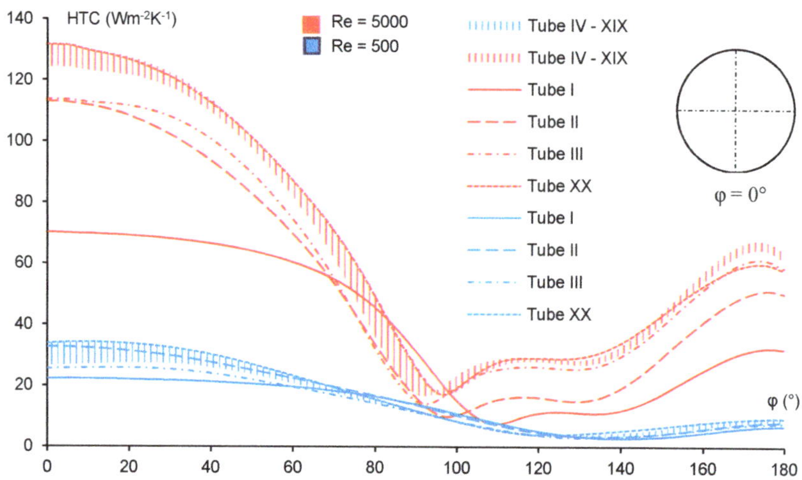

Figure 12. The local heat transfer coefficients of the staggered tube bank.

This paper deals with an approach where a constant tube wall temperature is considered. According to Žukauskas [20], when a constant heat flux is set on the tube wall, the temperature is variable and the average Nusselt number is generally higher. The difference increases with the increase in the Reynolds number.

When considering gravity, it is proved that at lower Reynolds numbers, gravity has a significant effect on the local Nusselt numbers for a single tube. The higher the Reynolds number, the lower the Nusselt number difference noticed. Since gravity is ubiquitous in real life, it was considered in the paper as well.

4. Conclusions

The reference temperature determination is required when computing the Nusselt number based on the heat transfer coefficient. For some geometries, the reference temperature is unambiguously determined. The tube bank does not have a unique location and authors use different ways to determine the temperature, whereby a comparison becomes difficult. This paper considers several temperature locations in a tube bank subdomain and compares the Nusselt numbers with the most used correlating equations. On the basis of the Žukauskas correlating equations, a unique reference temperature for both inline and staggered tube bank arrangements is determined with two Reynolds number ranges.

Concerning the inline tube bank, the temperature in the half of the subdomain inlet is the most appropriate reference temperature for the Reynolds number in the range of 10^2 and 10^3. Over this interval, a combination of the temperature in the bound of the subdomain inlet, the temperature in the free stream of the subdomain center, and the wall temperature should be used. When the staggered tube bank is considered, the mass weighted average temperature at the subdomain inlet is recommended for $Re = 10^2$–10^3, while the temperature in the free stream of the subdomain center should be used over this interval. The mentioned approach provides more accurate numerical results that are comparable with those obtained by the experiments.

Author Contributions: Conceptualization, Z.B.; methodology, S.K.; software, S.K.; validation, S.K.; formal analysis, S.K.; investigation, Z.B.; resources, Z.B.; data curation, S.K.; writing—original draft preparation, S.K.; writing—review and editing, Z.B.; visualization, S.K.; supervision, S.K.; project administration, Z.B.; funding acquisition, Z.B. All authors have read and agreed to the published version of the manuscript.

Funding: The paper has been written on the basis of the research intention and solution of the research grant project VEGA no. 1/0086/18 "Researching Temperature Fields in a Set of Shaped Heat Transfer Surfaces" and "Progressive Research into Utility Properties of Materials and Products Based on Wood (LignoPro)", ITMS 313011T720, supported by the Operational Programme Integrated Infrastructure (OPII), funded by the ERDF.

Institutional Review Board Statement: Not applicable.

Informed Consent Statement: Not applicable.

Data Availability Statement: Not applicable.

Conflicts of Interest: The authors declare no conflict of interest.

References

1. Liu, X.; Wang, M.; Liu, H.; Chen, W.; Qian, S. Numerical Analysis on Heat Transfer Enhancement of Wavy Fin-Tube Heat Exchangers for Air-Conditioning Applications. *Appl. Therm. Eng.* **2021**, *199*, 117597. [CrossRef]
2. Wang, Y.; Gu, X.; Jin, Z.; Wang, K. Characteristics of Heat Transfer for Tube Banks in Crossflow and its Relation with that in Shell-and-Tube Heat Exchangers. *Int. J. Heat Mass Transf.* **2016**, *93*, 584–594. [CrossRef]
3. Amatachaya, P.; Krittacom, B. Combustion Mechanism of Gas Porous Burner Installed an In-Line Tube-Bank Heat Exchanger. *Energy Procedia* **2017**, *138*, 50–55. [CrossRef]
4. Gu, L.D.; Min, J.C. Airside Thermal-Hydraulic Characteristics for Tube Bank Heat Exchangers Used to Cool Compressor Bleed Air in an Aero Engine. *Appl. Therm. Eng.* **2018**, *141*, 939–947. [CrossRef]
5. Wang, Y.; Tian, H.; Shu, G.; Yu, G.; Ma, X.; Li, X. Simulation and Optimization of Metal-foam Tube Banks for Heat Transfer Enhancement of Exhaust Heat Exchangers. *Energy Procedia* **2017**, *142*, 3863–3869. [CrossRef]
6. Kang, H.C.; Eoh, J.H.; Cha, J.E.; Kim, S.O. Numerical Study on Pressure Drop and Heat Transfer for Designing Sodium-to-Air Heat Exchanger Tube Banks on Advanced Sodium-Cooled Fast Reactor. *Nucl. Eng. Des.* **2013**, *254*, 5–15. [CrossRef]
7. Barnoon, P.; Toghraie, D.; Mehmandoust, B.; Fazilati, M.A.; Eftekhari, S.A. Comprehensive Study on Hydrogen Production via Propane Steam Reforming Inside a Reactor. *Energy Rep.* **2021**, *7*, 929–941. [CrossRef]
8. El-Shorbagy, M.A.; Eslami, F.; Ibrahim, M.; Barnoon, P.; Xia, W.F.; Toghraie, D. Numerical Investigation of Mixed Convection of Nanofluid Flow in a Trapezoidal Channel with Different Aspect Ratios in the Presence of Porous Medium. *Case Stud. Therm. Eng.* **2021**, *25*, 100977. [CrossRef]
9. Bairi, A. Porous Materials Saturated with Water-Copper Nanofluid for Heat Transfer Improvement between Vertical Concentric Cones. *Int. Commun. Heat Mass Transf.* **2021**, *126*, 105439. [CrossRef]
10. Miles, A.; Bessaih, R. Heat Transfer and Entropy Generation Analysis of Three-Dimensional Nanofluids Flow in a Cylindrical Annulus Filled with Porous Media. *Int. Commun. Heat Mass Transf.* **2021**, *124*, 105240. [CrossRef]
11. Grimison, E.D. Correlation and Utilization of New Data on Flow Resistance and Heat Transfer for Cross Flow of Gases over Tube Banks. *Trans. ASME* **1937**, *59*, 583–594.
12. Pierson, O.L. Experimental Investigation of the Influence of Tube Arrangement on Convection Heat Transfer and Flow Resistance in Cross Flow of Gases over Tube Banks. *Trans. ASME* **1937**, *59*, 563–572.

13. Huge, E.C. Experimental Investigation of Effects of Equipment Size on Connection Heat Transfer and Flow Resistance in Cross Flow of Gases over Tube Banks. *Trans. ASME* **1937**, *59*, 573–581.
14. Brevoort, M.J.; Tifford, A.N. *An Experimental Investigation of Flow across Tube Banks*; ARR (WR L-232); Langley Aeronautical Laboratory: Washington, WA, USA, 1942; pp. 1–10.
15. McAdams, W.H. *Heat Transmission*; McGraw-Hill Company: New York, NY, USA, 1942.
16. Schmidt, F.; Werner, K. *Heat Transfer over the Circumference of a Heated Cylinder in Transverse Flow*; Technical Memo (1050); National Advisory Committee for Aeronautics: Washington, WA, USA, 1943.
17. Kays, W.M.; Lo, R.K. *Basic Heat Transfer and Flow Friction Design Data for Gas Flow Normal to Banks of Staggered Tubes*; Stanford University: Stanford, CA, USA, 1952; pp. 1–74.
18. Welch, C.P.; Fairchild, H.N. Individual Row Heat Transfer in a Crossflow: In-line Tube Bank. *Trans. ASME* **1964**, *86*, 143–148. [CrossRef]
19. Samoshka, P.S.; Makaryavichyus, V.I.; Shlachyauskas, A.A.; Zhyughda, I.I.; Žukauskas, A. Heat Transfer and Pressure Drop for Closely Spaced Tube Banks in Water Flow. *Int. Chem. Eng.* **1968**, *18*, 388–392.
20. Žukauskas, A. Heat Transfer from Tubes in Crossflow. *Adv. Heat Transf.* **1972**, *8*, 93–160.
21. Žukauskas, A.; Ulinskas, R.V. Heat Transfer Efficiency of Tube Bundles in Crossflow at Critical Reynolds Numbers. *Heat Transf. -Sov. Res.* **1978**, *10*, 9–15.
22. Žukauskas, A.; Ulinskas, R.V. *Heat Transfer in Tube Banks in Crossflow*; Hemisphere: New York, NY, USA, 1988.
23. Hausen, H. *Heat Transfer in Counterflow Parallel Flow and Cross Flow*; McGraw-Hill: New York, NY, USA, 1983.
24. Ramezanpour, A.; Mirzaee, I.; Rahmani, R.; Shirvani, H. Numerical Study of Staggered Tube Bundle in Turbulent Cross Flow for an Optimum Arrangement. *Int. J. Heat Exch.* **2006**, *7*, 37–56.
25. Khan, W.A.; Culham, J.R.; Yovanovich, M.M. Analytical Model for Convection Heat Transfer from Tube Banks. *J. Thermophys. Heat Transf.* **2006**, *20*, 720–727. [CrossRef]
26. Khan, W.A.; Culham, J.R.; Yovanovich, M.M. Convection Heat Transfer from Tube Banks in Crossflow: Analytical Approach. *Int. J. Heat Mass Transf.* **2006**, *49*, 4831–4838. [CrossRef]
27. Haider, M.J.; Danish, S.N.; Khan, W.A.; Mehdi, S.U.; Abbasi, B.A. Heat Transfer and Fluid Flow over Circular Cylinders in Cross Flow. *NUST J. Eng. Sci.* **2010**, *3*, 67–77.
28. Kim, T. Effect of Longitudinal Pitch on Convective Heat Transfer in Crossflow over In-Line Tube Banks. *Ann. Nucl. Energy* **2013**, *57*, 209–215. [CrossRef]
29. Yilmaz, A.; Yilmaz, T. Optimum Design of Cross-Flow In-Line Tube Banks at Constant Wall Temperature. *Heat Transf. Eng.* **2015**, *37*, 523–534. [CrossRef]
30. Yilmaz, A.; Erdinc, M.T.; Yilmaz, T. Optimization of Crossflow Staggered Tube Banks for Prescribed Pressure Loss and Effectiveness. *J. Thermophys. Heat Transf.* **2017**, *31*, 4. [CrossRef]
31. Niemelä, N.P.; Mikkonen, A.; Lampio, K.; Konttinen, J. Computational Fluid Dynamics Based Approach for Predicting Heat Transfer and Flow Characteristics of Inline Tube Banks with Large Transverse Spacing. *Heat Transf. Eng.* **2021**, *42*, 270–281. [CrossRef]
32. Córcoles, J.I.; Marín-Alarcón, E.; Almendros-Ibánez, J.A. Heat Transfer Performance of Fruit Juice in a Heat Exchanger Tube Using Numerical Simulations. *Appl. Sci.* **2020**, *10*, 648. [CrossRef]
33. Beale, S.B. Fluid Flow and Heat Transfer in Tube Banks. Ph.D. Thesis, University of London, London, UK, 1993.
34. Ge, Y.; Lin, Y.; Tao, S.; He, Q.; Chen, B.; Huang, S.M. Shape Optimization for a Tube Bank Based on the Numerical Simulation and Multi-Objective Genetic Algorithm. *Int. J. Therm. Sci.* **2021**, *161*, 106787. [CrossRef]
35. Gnielinski, V. *Wärmeübergang bei Querströmung durch einzelne Rohrreihen und durch Rohrbündel*; Springer: Berlin/Heidelberg, Germany, 1974.
36. Bergelin, O.P.; Leighton, M.D.; Lafferty, W.L.; Pigford, R.L. *Heat Transfer and Pressure Drop during Viscous and Turbulent Flow across Baffled and Unbaffled Tube Banks*; Engineering Experiment Station Bulletin 4; University of Delaware: Newark, DE, USA, 1958.
37. Le Feuvre, R.F. Laminar and Turbulent Forced Convection Processes through In-Line Tube Banks. Ph.D. Thesis, University of London, London, UK, 1973.
38. Massey, T.H. The Prediction of Flow and Heat Transfer in Banks of Tubes in Cross-Flow. Ph.D. Thesis, Central Electricity Research Laboratories, Surrey, UK, 1976.
39. Castro, L.L.; Aranda, A.; Urquiza, G. Analysis of Heat Transfer in an Experimental Heat Exchanger Using Numerical Simulation. In *Numerical Simulation—From Brain Imaging to Turbulent Flows*; IntechOpen: London, UK, 2016. [CrossRef]
40. Mangrulkar, C.h.K.; Dhoble, A.S.; Pant, P.K.; Kumar, N.; Gupta, A.; Chamoli, S. Thermal Performance Escalation of Cross Flow Heat Exchanger using In-Line Elliptical Tubes. *Exp. Heat Transf.* **2019**, *33*, 587–612. [CrossRef]
41. Muzaffar, A.; Cheema, T.A.; Abbas, A.; Tayyab, M.; Ilyas, M.; Park, C.h.W. Performance analysis of liquified petroleum gas (LPG) driven half-cycle air conditioning system. *Heat Mass Transf.* **2020**, *56*, 3177–3197. [CrossRef]
42. Unger, S.; Beyer, M.; Pietruske, H.; Szalinski, L.; Hampel, U. Natural Convection Heat Transfer Performance of Additively Manufactured Tube Bundle Heat Exchangers with Novel Fin Design. *Heat Mass Transf.* **2021**, *57*, 1193–1203. [CrossRef]
43. Xu, L.; Chen, H.; Xi, L.; Xiong, Y.; Gao, J.; Li, Y. Flow and Heat Transfer Characteristics of a Staggered Array of Kagome Lattice Structures in Rectangular Channels. *Heat Mass Transf.* **2021**. [CrossRef]

44. Wang, L.C.; Su, M.; Hu, W.L.; Lin, Z.M.; Wang, L.B.; Wang, Y. The Characteristic Temperature in the Definition of Heat Transfer Coefficient on the Fin Side Surface in Tube Bank Fin Heat Exchanger. *Numer. Heat Transf. A* **2011**, *60*, 848–866. [CrossRef]
45. Cengel, Y.A.; Ghajar, A.J. *Heat and Mass Transfer*; McGraw-Hill Education: New York, NY, USA, 2015.
46. Yuan, W.; Fang, G.; Zhang, X.; Tang, Y.; Wan, Z.; Zhang, S. Heat Transfer and Friction Characteristics of Turbulent Flow through a Circular Tube with Ball Turbulators. *Appl. Sci.* **2018**, *8*, 776. [CrossRef]
47. Engineering Sciences Data Unit. Convective Heat Transfer during Crossflow of Fluids Over Plain Tube Banks, ESDU Data Item No. 7303J. 1973. Available online: https://www.thermopedia.com/content/1212 (accessed on 7 October 2021).

applied sciences

Article

Influence of Blade Wrap Angle on the Hydrodynamic Radial Force of Single Blade Centrifugal Pump

Linwei Tan *, Yongfei Yang *, Weidong Shi *, Cheng Chen and Zhanshan Xie

School of Mechanical Engineering, Nantong University, Nantong 226019, China; cc1216734906@163.com (C.C.); xiezs@ntu.edu.cn (Z.X.)
* Correspondence: tanlinwei@ntu.edu.cn (L.T.); yyf2020@ntu.edu.cn (Y.Y.); wdshi@ujs.edu.cn (W.S.)

Abstract: To investigate the effect of blade wrap angle on the hydrodynamic radial force of a single blade centrifugal pump, numerical simulation is conducted on the pumps with different blade wrap angles. The effect of the wrap angle on the external characteristics and the radial force of a single blade centrifugal pump was analyzed according to the simulation result. It is found that, with the increase of the blade wrap angle, the head and efficiency of the single blade centrifugal pump are improved, the H-Q curve becomes steeper, and the efficiency also increased gradually, while the high-efficiency area is narrowed. The blade wrap angle has a great effect on the radial force of the single blade centrifugal pump. When the blade wrap angle is less than 360°, the horizontal component of the radial force is negative and the value is reduced with the increase of the wrap angle of the blade. When the wrap angle is larger than 360°, the horizontal component of the radial force is positive and the value increases with the increase of the wrap angle. Under part-loading conditions, the radial force of the single blade pump is significantly reduced with the increase of the blade wrap angle. When the wrap angle is smaller than 360°, the radial force decreases with the flow rate increase. In the condition that the wrap angle is larger than 360°, the radial force increases with the flow rate increase.

Keywords: blade wrap angle; hydrodynamic radial force; single blade centrifugal pump; numerical simulation

Citation: Tan, L.; Yang, Y.; Shi, W.; Chen, C.; Xie, Z. Influence of Blade Wrap Angle on the Hydrodynamic Radial Force of Single Blade Centrifugal Pump. *Appl. Sci.* **2021**, *11*, 9052. https://doi.org/10.3390/app11199052

Academic Editor: Jesús María Blanco

Received: 2 September 2021
Accepted: 26 September 2021
Published: 28 September 2021

1. Introduction

A single blade centrifugal pump has a good non-clogging performance, which is widely used in the process of wastewater treatment [1]. To improve the ability of the flow channel to pass particles and strip impurities, the impeller is designed with only one blade that has a very large wrap angle. Such a kind of non-axisymmetric structure brings different kinds of problems to the single blade centrifugal pump, such as larger radial force, shock, and vibrations than the traditional centrifugal pumps.

With the development of high-speed and high-capacity pumps, the safety and reliability of the centrifugal pump have attracted attention from engineers and researchers [2]. Unstable problems like shock and vibrations are usually caused by the unsteady flow characteristics in the pump, which is extensively investigated till now. Stepanoff and Biheller [3,4] first proposed the radial force empirical formula, which estimates the radial force experienced by the impeller based on the impeller geometric parameters, flow rate and head, but the empirical coefficient has certain limitations. Brennen [5–8] comprehensively and systematically tested the radial force of NASA high-speed liquid oxygen and liquid hydrogen turbo pumps in the United States. The effect of impeller eccentricity, front and rear pump cavity leakage, and vortex frequency on the radial force were analyzed to establish the mathematical model of the radial force. Guelich [9] summarized the reasons for the radial force of the pump in his review: the uneven distribution of the circumferential pressure of the impeller, the radial force generated by the labyrinth seal, the dynamic and static interference effect of the impeller guide vane (volute), the steady radial force, the specific speed of the pump, the type of impeller or guide vane, the geometric parameters of the impeller and the operating conditions are related. Boehning [10] compared the effects of single volute, annular volute, and double volute on the radial force of the blood pump using a combination of numerical simulation and experiment. At the designed flow conditions, the radial forces are equivalent in the case of single volute and annular volute, and all have large radial forces, while the double volutes have almost no radial forces. Alemi [11] analyzed the influence of different volute structures on the radial force based on numerical simulation. The results showed that when the volute diaphragm was at 270°, the radial force was the smallest in all flow rate conditions.

Due to the asymmetric structure of the impeller, the radial force problem of single blade centrifugal pumps is particularly prominent, which makes the pumps less stable and the vibration-induced noise is stronger. Benra [12–15] has carried out a lot of research on the flow characteristics and hydraulic induced vibration in single-channel centrifugal pumps. Aoki [16] measured the transient pressure distribution of an open single-blade centrifugal impeller and obtained the dynamic and static radial forces of the impeller. Nishi [17–19] conducted a lot of research on the radial force of the single blade centrifugal pump by combining numerical simulation and experiment. The radial force on the single blade impeller was measured by installing a force measuring ring at the bearing. The influence of the blade outlet angle and blade outlet width on the radial force was investigated. It was found that to increase the blade outlet width can reduce the averaged value of the radial force at part-loading conditions. The radial force is divided into the inertial term, momentum term, and pressure term. The calculation results are basically consistent with the numerical calculation pressure integral results.

Since the single blade centrifugal pump has only one blade, to select reasonable parameters for the impeller during the design process is of great importance. Tan analyzed the influence of blade wrap angle on a single-channel pump performance [20]. Chen analyzed the effect of blade inlet angle on a single blade centrifugal pump performance [21]. The wrap angle is one of the main parameters for the design of centrifugal pumps. However, research concerning the effect of the blade wrap angle on the hydrodynamic radial force and the reliability of single blade centrifugal pump is lacking. In the current research, unsteady simulation is conducted for single blade centrifugal pumps with different blade wrap angles, and the effects of these pumps are compared to reveal the relationship between the blade wrap angle and the pump performance.

2. Pump Model and Simulation Method

2.1. Pump Model

The pump model in this study is a single blade centrifugal pump which is the same as the author's early paper [22]. The main design parameters are Q_d = 20 m^3/h, H = 11 m, rotational speed n = 2940 r/min, and the main geometry parameters are shown as Table 1. The impeller was 3D designed by Bladegen based on the main parameters, and the design process of single blade centrifugal impeller by Bladegen is shown in Figure 1. The blade wrap angle is an important parameter of a single blade centrifugal pump. Properly increasing the blade wrap angle can enhance the restriction of the impeller to the fluid, but an excessively large blade wrap angle will result in increased friction loss. In the current research, impellers with five different wrap angles are investigated, the values of the wrap angles are 300°, 330°, 360°, 390°, and 420°. Other parameters for the impellers are kept consistent, which is shown in Table 1.

Table 1. Main geometric parameters of the pump.

Main Parameters	Value
Inlet diameter of impeller, D_j (mm)	45
Outlet diameter of impeller, D_2 (mm)	125
Outlet width of impeller, b_2 (mm)	30
Blade wrap angle, φ (°)	360
Inlet diameter of volute, D_3 (mm)	135
Inlet width of volute, b_3 (mm)	46
Blade outlet angle, β_2 (°)	18

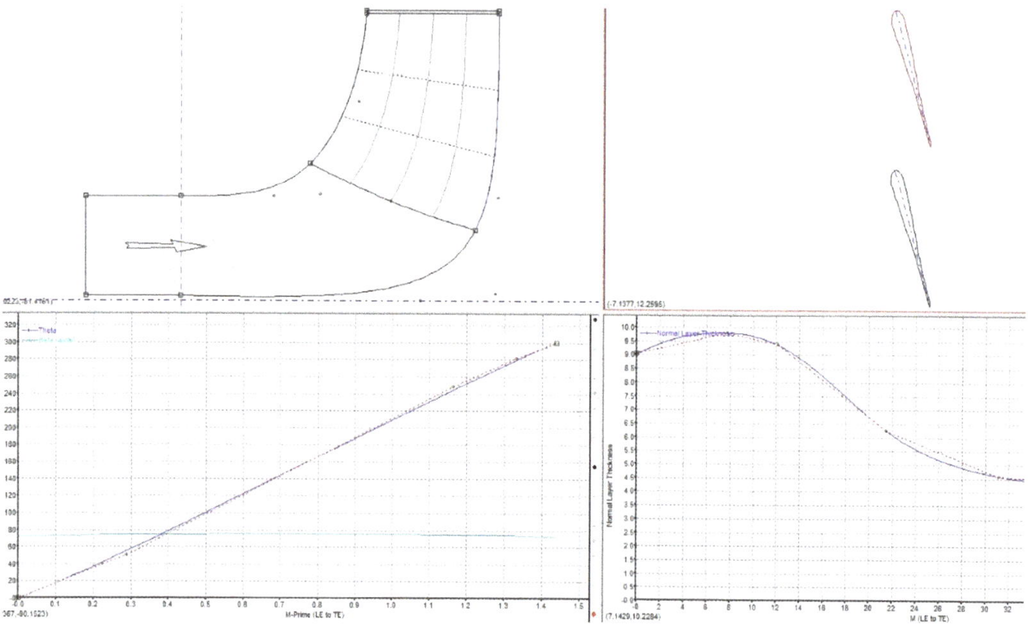

Figure 1. Design process of single blade centrifugal impeller by Bladegen.

2.2. Simulation Method

The simulation method is the same as the author's early paper, which is validated by experiment [22]. The ICEM code was used to mesh the model, and tetrahedral mesh was used for the impeller. To improve the grid quality, we refined the wall boundary layer. Meshing results of the impellers are illustrated in Figure 2. The other calculation domains adopted the hexahedral structure mesh the same as the reference [22].

As the flow in the centrifugal pump has a high Reynolds number, the numerical simulation was adapted from the standard *k*-ε turbulence model [23–26]. ANSYS CFX 14.5 is used for numerical calculation, the coupling of velocity and pressure adopts the SIMPLEC algorithm, and the convection term adopts high-order format [27]. Using 25 °C water as the calculation medium, the solid wall is selected as a nonslip wall surface. Since the impeller and volute are castings, the surface roughness is set as 50 μm. The inlet is set to the total pressure inlet, and the outlet is set to the mass outlet. By setting different mass flows, the external characteristic curve of the pump can be obtained. A multi-coordinate reference system is adopted, the impeller is set to the rotating domain, the static domain is adopted for the others, and the interface between the dynamic and static domains is set to the frozen rotor. Considering the calculation time and accuracy, the convergence accuracy is set to 10^{-4}.

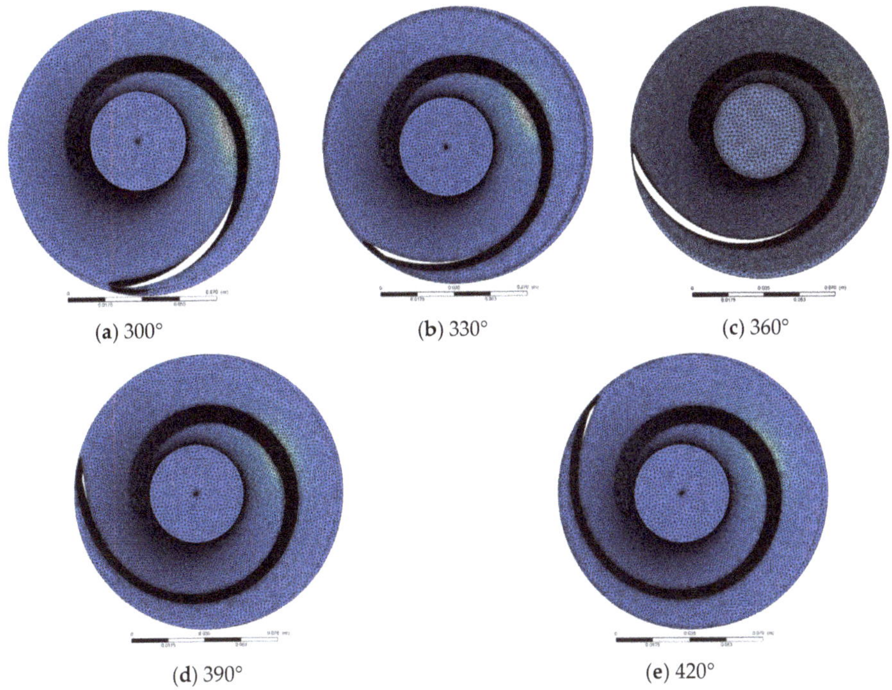

Figure 2. Mesh of different impellers.

2.3. Validation

The experiment of a single blade centrifugal pump with 360° blade wrap angle was conducted [22]. Pump performance experimental setup and distribution of pressure sensors are shown in Figures 3 and 4. The experimental device consists of the test pump, connecting pipes, valves, electromagnetic flowmeter and other components. Four pressure taps were located in the spiral volute wall. Pressure taps were instrumented with fast response pressure sensors (Figure 4), which provided absolute pressure values with an uncertainty of less than 0.1% according to the manufacturer's data.

Figure 3. Experimental device: (1) valve in the outlet, (2) flowmeter, (3) valve in the inlet, (4) pressure transducer in the outlet, (5) pressure transducer in the inlet, (6) pump.

Figure 4. Pressure sensors distribution.

The pump performance curves of single blade centrifugal pump with 360° blade wrap angle were acquired by turn off or turn on the valve in the outlet. Figure 5 presents the results of CFD and experiment. The predicted power of CFD is hydraulic power, whereas the result of the experiment is total power. In order to compare with CFD results, based on the empirical coefficients of different power levels in the laboratory, the measured total efficiency is transformed into hydraulic efficiency. The experimental heads are smaller than numerical results, and it is less than 1 m. It can be observed that the CFD results are in good agreement with the experimental results.

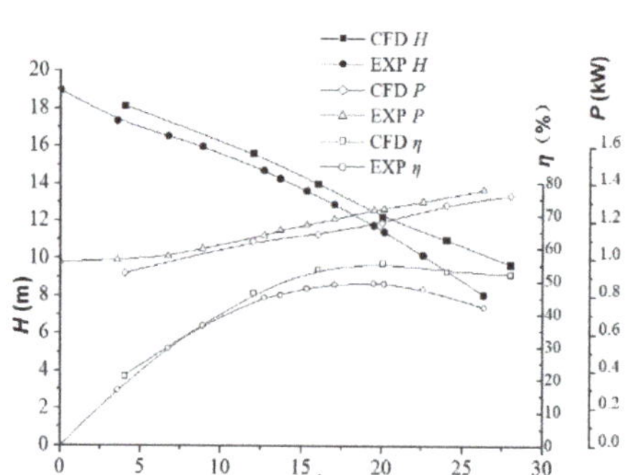

Figure 5. Pump performance.

The pressure distribution of model 1 is presented in Figure 6 [22]. The CFD predicted pressures were similar to those obtained by experiment. The time history of the pressure in the numerical simulation approximately overlapped with the test results. The qualitative agreement of the pressure obtained by CFD numerical calculation and experiment was extremely high; and the CFD results show a reasonable agreement with the test data. Thus, numerical calculation results can be considered reliable.

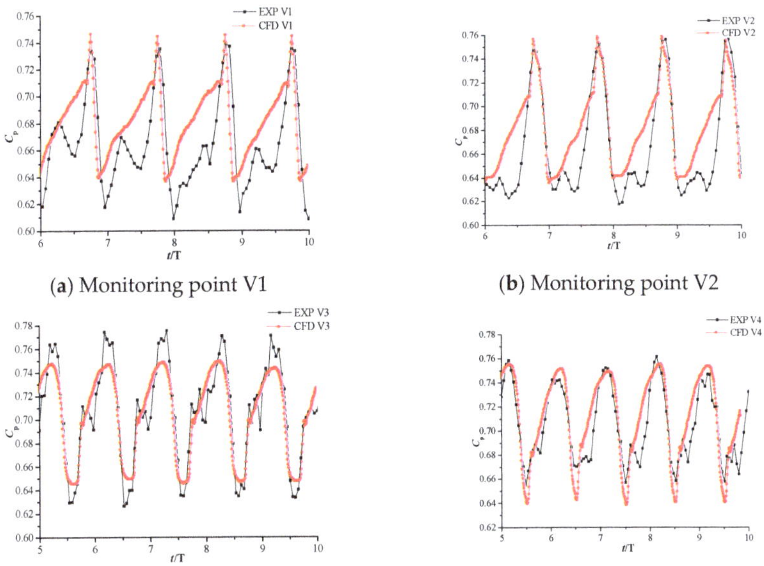

(**a**) Monitoring point V1 (**b**) Monitoring point V2

(**c**) Monitoring point V3 (**d**) Monitoring point V4

Figure 6. Comparison of pressure distribution in model 1 at design flowrate.

3. Results and Discussion

3.1. Influence of Blade Wrap Angle on the External Characteristic

To analyze the effect of the wrap angle on the hydrodynamic radial force, five impellers were designed. The external characteristic curves of single blade centrifugal pumps with different blade wrap angles are shown in Figure 7. It can be seen from the figure that, as the blade wrap angle increases, the pump head tends to increase, and the *H*-*Q* curve becomes steeper, the pump efficiency increases gradually, but the high-efficiency area is narrowed. The *P*-*Q* curve becomes gentle with the increase of blade wrap angle. Generally, the head and efficiency of the single blade centrifugal pump are improved with the increase of blade wrap angle.

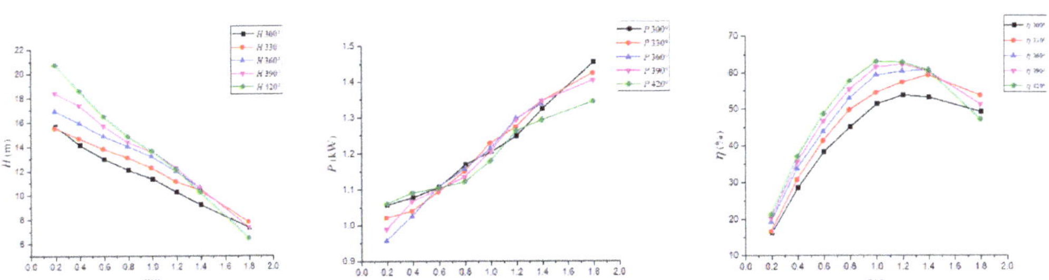

(**a**) *H*-*Q* (**b**) *P*-*Q* (**c**) *η*-*Q*

Figure 7. External characteristics of single blade centrifugal pump with different wrap angles.

The turbulent dissipation rate cloud diagram of single blade centrifugal pump with different blade wrap angles at designed flow conditions are shown in Figure 8. It can be seen from the figure that the turbulence dissipation rate of single blade centrifugal pumps with blade wrap angle less than 360° is significantly higher. For other impellers, the dissipation rate changes no longer significantly after the wrap angle reaches 360°.

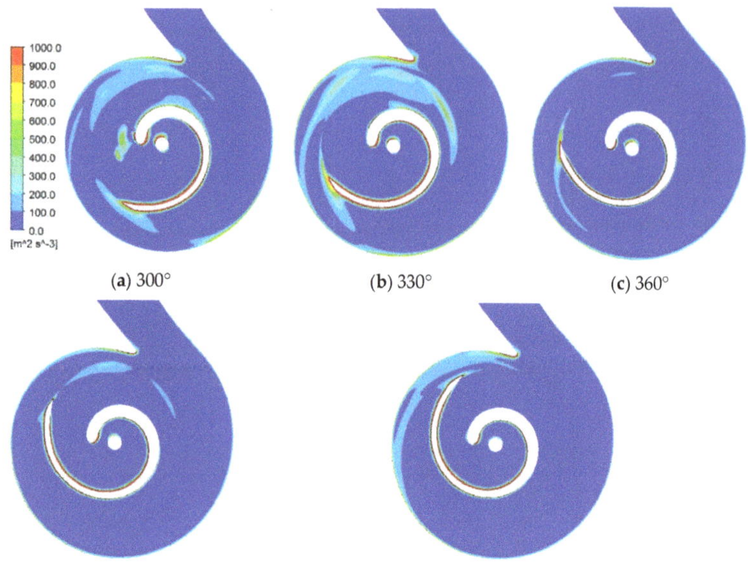

(**a**) 300° (**b**) 330° (**c**) 360°

(**d**) 390° (**e**) 420°

Figure 8. Turbulence dissipation rate of single blade centrifugal pump with different wrap angle.

The turbulent dissipation losses in a single blade impeller with different blade wrap angles are shown in Figure 9a. It can be seen from the figure that the dissipation losses show a decreasing trend as the blade wrap angle increases, but after the wrap angle reaches 420°, the loss of dissipation has increased. The friction loss of the single blade centrifugal impeller with different wrap angles is shown in Figure 9b. As can be seen from the figure, under the designed flow rate conditions, the flow of the impeller continuously improves with the increase of the blade wrap angle, so the friction losses decrease with the wrap angle increase. However, under high flow rate conditions, because the friction surface increases with the increase of the wrap angle under high flow velocity, the friction loss increases with the increase of the wrap angle. Therefore, the efficiency of the single blade centrifugal pumps, which have a large wrap angle, decreases at large flow conditions, and its Q-η curve in the high-efficiency area becomes narrow.

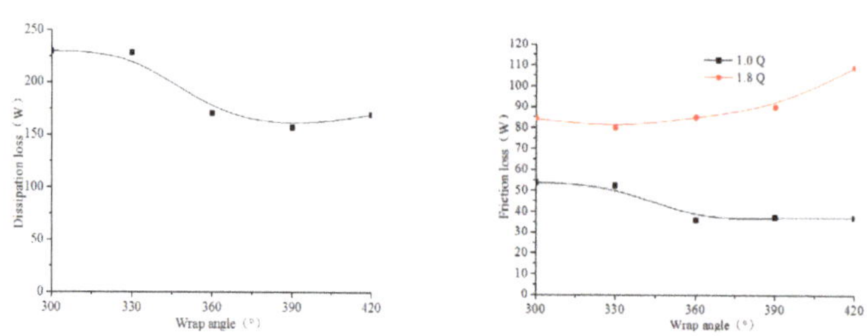

(**a**) Turbulence dissipation losses (**b**) Disc friction losses

Figure 9. Energy loess of single blade centrifugal pump with different wrap angles.

3.2. *Influence of Blade Wrap Angle on Radial Force*

The blade wrap angle affects the pressure circular distribution in the impeller. As a result, the wrap angle is an important factor that affects the radial force of the single blade centrifugal pump. In this section, the effect of wrap angle on the radial force of the single blade centrifugal pump is investigated based on unsteady simulation. The unsteady simulation is validated by experiment and the simulation results agree well with the tested values [22]. In order to calculate the radial force, the impeller was divided as: FS, the front shroud; BS, back shroud; B, blade [28]. The Cartesian coordinates are used to specify the radial force. It is specified that the radial force in the positive X direction and the positive Y direction is positive; the corresponding forces in the negative X direction and the negative Y direction are negative, as shown in Figure 10.

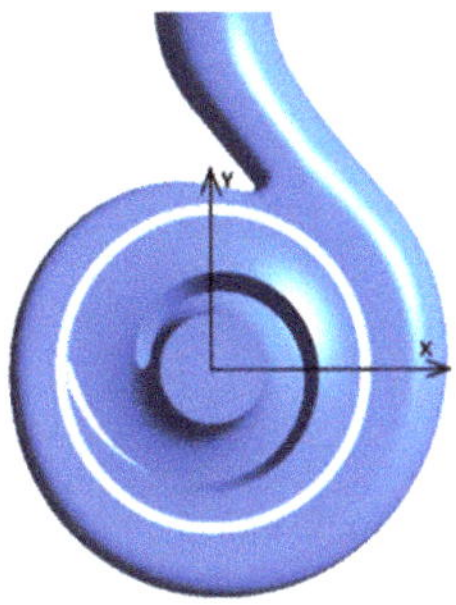

Figure 10. Coordinate diagram of force.

Figure 11 shows the time domain diagram of the radial force on the impeller of a single blade centrifugal pump with different blade wrap angles under design flow rate. It can be found from the figure, the radial force fluctuates periodically with the rotation of the impeller, and the periodicity is consistent. The radial force on the back shroud is small, and it does not change significantly with the variation of the wrap angle. The amplitude and direction of the x component of the radial force changes with the change of the blade wrap angle. When the blade wrap angle is 300°, 330° and 360°, the x component of the radial force is negative and the value decreases with the increase of the blade wrap angle. When the blade angle is larger than 390°, the x component of the radial force turns to positive and the value increases with the increase of the wrap angle. When the wrap angle is 360° and 390°, the y component of the radial force is the largest. When the wrap angle is 300° the y component of the radial force is the lowest.

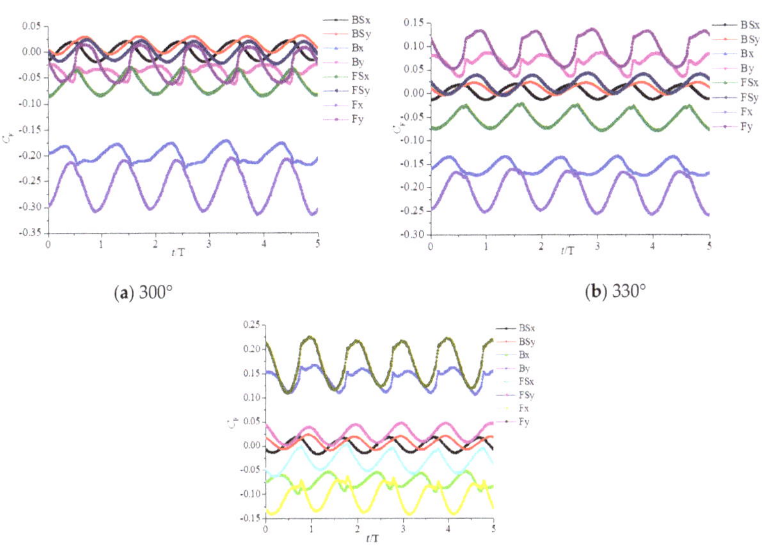

Figure 11. *Cont.*

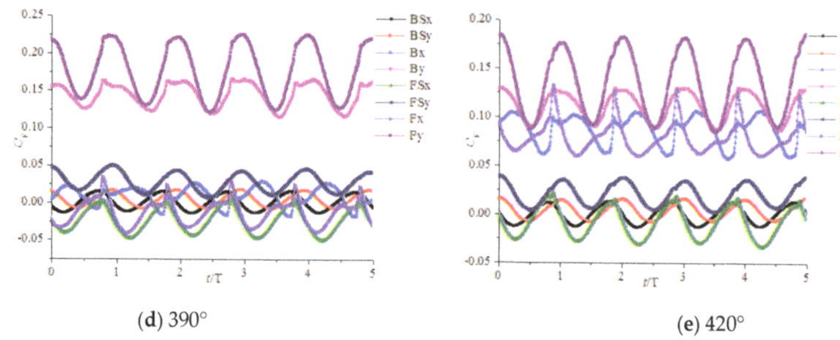

Figure 11. Radial force of a single centrifugal impeller with different wrap angles under the designed flow rate condition.

Figure 12 shows the radial force F_r curve of the impeller with different blade wrap angles when the impeller rotates. The calculation results are adopted the dimensionless coefficient C_F.

$$C_F = \frac{2F}{\rho u_2{}^2 D_2 b_2} \tag{1}$$

where C_F is the dimensionless coefficient of the force, F is the radial force, ρ is the density, u_2 is the circumferential velocity of the impeller outlet, D_2 is the impeller outer diameter, b_2 is the impeller outlet width.

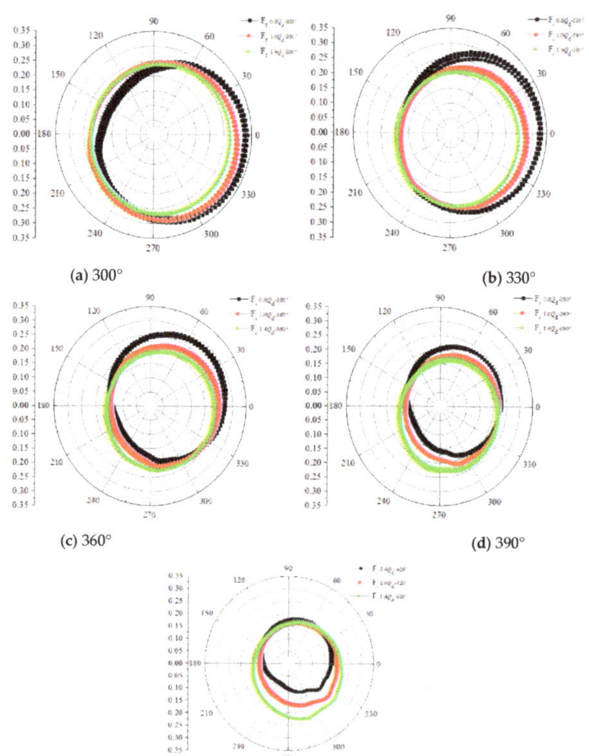

Figure 12. Radial forces of the impellers with different blade wrap angles according to the rotation of the impeller.

It can be seen from the figure, when the blade wrap angle is less than 360°, the radial force is significantly greater than that of the impeller with other wrap angles. Under low flow conditions, the radial force of the impeller significantly decreases continuously with the increase of the blade wrap angle.

Figures 13 and 14 show the vector and pressure contour of the single blade centrifugal pump with different wrap angles. It can be seen from the vector diagram that the flow field is disturbed when the single blade centrifugal pump is working under part-loading conditions, there is a sudden change in the streamline, and there is an obvious low-speed vortex area in the impeller which means flow separation occurred. When the wrap angle of the impeller blade is increased, the flow field inside the pump is improved. From the pressure contour, it can be found that, with the increase of the wrap angle, the pressure distribution in the circular direction becomes relatively uniform. Therefore, increasing the wrap angle of the blade is a useful way to reduce the radial force under part-loading conditions. Under the designed flow rate condition or larger flow rate conditions, the radial force of the impeller also decreases with the increase of the wrap angle of the blade, and the effect is weakened when the wrap angle is larger than 360°.

(**a**) 300° (**b**) 330° (**c**) 360° (**d**) 390° (**e**) 420°

Figure 13. Velocity vector of the pump under part-loading conditions (0.6 Q_d).

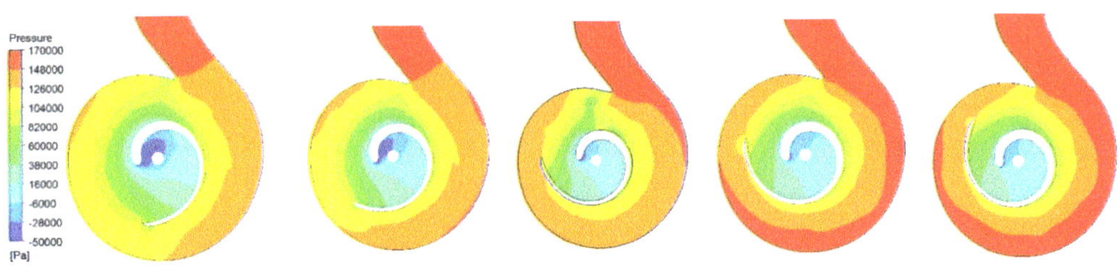

(**a**) 300° (**b**) 330° (**c**) 360° (**d**) 390° (**e**) 420°

Figure 14. Pressure distribution in the pump under part-loading conditions (0.6 Q_d).

Figures 15–18 show the velocity vector and the pressure contour under the designed flow rate and larger flow rate conditions. Under these working conditions, the flow field in the pump is good, while for the impellers with 300° and 330° wrap angles, flow separation still happens. The pressure distribution in the circular direction is not axisymmetric, while the axisymmetric characteristic is improved when the wrap angle is increased. Comparing the performance of one impeller under different working conditions, it can be found that when the wrap angle is smaller than 360° the radial force decreases with the flow rate increase. When the wrap angle is 420°, the radial force is smaller under part-loading conditions, which increases with the flow rate increase. From the pressure cloud diagram of the pump with a blade wrap angle of 420°, it can be found that as the flow rate increases, a local low-pressure zone appears from the end of the spiral section of the volute to the outlet end, and the circumferential symmetry of the pressure distribution becomes worse resulting in an increase in the radial force.

(**a**) 300° (**b**) 330° (**c**) 360° (**d**) 390° (**e**) 420°

Figure 15. Velocity vector of the pump under the designed flow rate condition.

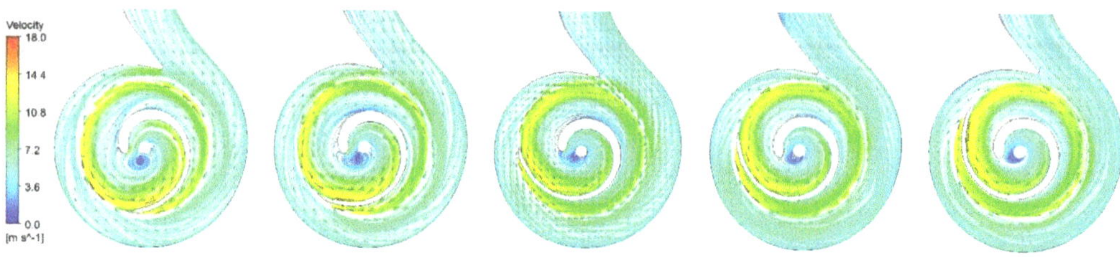

(**a**) 300° (**b**) 330° (**c**) 360° (**d**) 390° (**e**) 420°

Figure 16. Velocity vector of the pump large flow rate condition (1.4 Q_d).

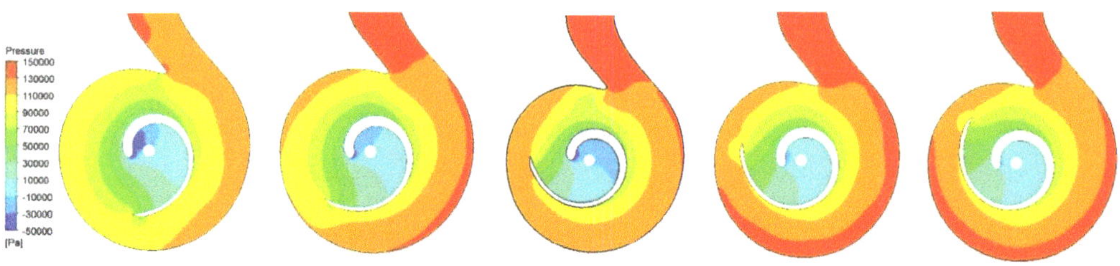

(**a**) 300° (**b**) 330° (**c**) 360° (**d**) 390° (**e**) 420°

Figure 17. Pressure distribution in the pump under designed flow rate conditions.

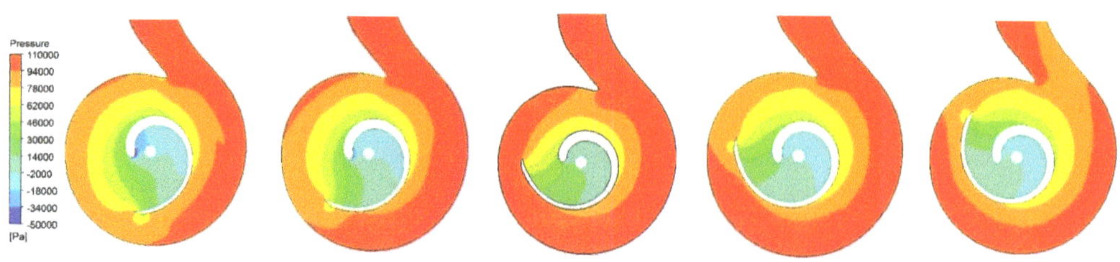

(**a**) 300° (**b**) 330° (**c**) 360° (**d**) 390° (**e**) 420°

Figure 18. Pressure distribution in the pump under large flow rate conditions (1.4 Q_d).

The main cause of the radial force of the single blade centrifugal pump is the uneven circumferential pressure distribution in the pump. Figure 19 shows the circumferential pressure distribution at the outlet of the impeller of the single blade centrifugal pump under designed flow rate conditions. From the figure, it can be clearly seen that the axisymmetric characteristic is improved with the increase of the blade wrap angle, while uneven pressure distribution is always found at the volute outlet, impeller inlet and impeller outlet.

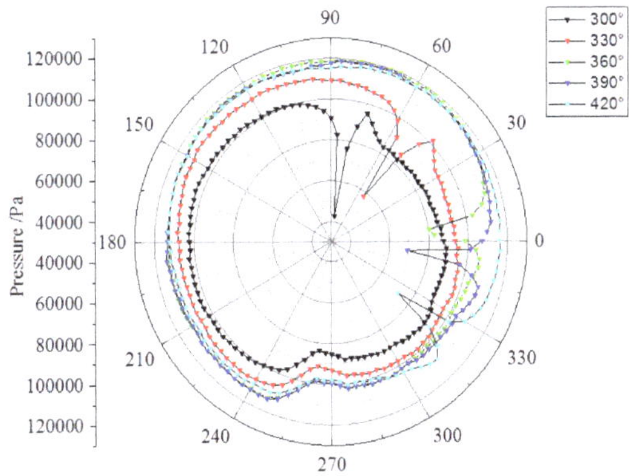

Figure 19. Pressure distribution at the impeller outlet under designed flow rate condition.

4. Conclusions

The effect of blade wrap angle on the performance of a single blade centrifugal pump is investigated based on unsteady numerical simulation. The external characteristics and radial force of single blade centrifugal pumps with five different wrap angles are simulated and analyzed. The research results are as follows:

(1) Generally, the head and efficiency of the single-blade centrifugal pump are improved with the increase of blade wrap angle. With the increase of the blade wrap angle, the pump head tends to increase, and the *H-Q* curve becomes steeper, the pump efficiency increases gradually, but the high-efficiency area is narrowed. The *P-Q* curve becomes gentle with the increase of blade wrap angle.

(2) The blade wrap angle has a significant impact on the radial force of the single blade centrifugal pump. When the blade wrap angle is less than 360°, the x component of the radial force is negative and the value is reduced with the increase of the wrap angle of the blade. When the wrap angle is larger than 360°, the x component of the radial force is positive and the value increases with the increase of the wrap angle. The y component of the impeller radial force has a maximum value when the wrap angle is 360° and 390°, and has a minimum value when the wrap angle is 300°.

(3) Under part-loading conditions, the radial force of the single blade pump is significantly reduced with the increase of the blade wrap angle. When the wrap angle is smaller than 360°, the radial force decreases with the flow rate increase. In the condition that the wrap angle is larger than 360°, the radial force increases with the flow rate increase.

(4) It suggested that the available blade wrap angle for single blade centrifugal pump should be between 360°and 420°, to achieve a better hydraulic performance and stable flow field.

Author Contributions: Data curation, L.T. and C.C.; Funding acquisition, W.S.; Investigation, Z.X.; Methodology, W.S.; Project administration, L.T.; Writing—original draft, L.T. and Y.Y. All authors have read and agreed to the published version of the manuscript.

Funding: This research was funded by the National Key Research and Development Project of China (no. 2019YFB 2005300), National High-Tech Ship Scientific Research Project of China (no. MIIT [2019] 360), the National Natural Science Foundation of China (No. 51979138), General project of natural science research in Colleges and universities of Jiangsu Province (No. 19KJB470029), and Jiangsu Water Conservancy Science and Technology Project (No. 2019038).

Conflicts of Interest: The authors declare no conflict of interest.

References

1. Benra, F.K.; Dohmen, H.; Sommer, M. Periodically unsteady flow in a single-blade centrifugal pump: Numerical and experimental results. In Proceedings of the ASME 2005 Fluids Engineering Division Summer Meeting, Hoston, TX, USA, 19–23 June 2005; American Society of Mechanical Engineers: New York, NY, USA, 2005; pp. 1223–1231.
2. Tang, S.; Yuan, S.; Zhu, Y. Convolutional Neural Network in Intelligent Fault Diagnosis toward Rotatory Machinery. *IEEE Access* **2020**, *8*, 86510–86519. [CrossRef]
3. Stepanoff, A. *Centrifugal and Axial Flow Pumps*; Van Chong Book Company: New York, NY, USA, 1950.
4. Biheller, H. Radial force on the impeller of centrifugal pumps with volute, semivolute, and fully concentric casings. *J. Eng. Power* **1965**, *87*, 319–322. [CrossRef]
5. Chamieh, D.S.; Acosta, A.J.; Brennen, C.E.; Caughey, T.K. Experimental measurements of hydrodynamic radial forces and stiffness matrices for a centrifugal pump-impeller. *J. Fluids Eng.* **1985**, *107*, 307–315. [CrossRef]
6. Adkins, D.R.; Brennen, C.E. Analyses of hydrodynamic radial forces on centrifugal pump impellers. *ASME Trans. J. Fluids Eng.* **1986**, *110*, 20–28. [CrossRef]
7. Yun, H.; Brennen, C.E. Effect of Swirl on Rotordynamic Forces Caused by Front Shroud Pump Leakage. *J. Fluids Eng.* **2002**, *124*, 1005–1010.
8. Brennen, C.E.; Acosta, A. Fluid-induced rotordynamic forces and instabilities. *Struct. Control Health Monit.* **2006**, *13*, 10–26. [CrossRef]
9. Guelich, J.; Jud, W.; Hughes, S.F. Review of Parameters Influencing Hydraulic Forces on Centrifugal Impellers. *Proc. Inst. Mech. Eng. Part A J. Power Energy* **1987**, *201*, 163–174. [CrossRef]
10. Boehning, F.; Timms, D.L.; Amaral, F.; Oliveira, L.; Graefe, R.; Hsu, P.L.; Schmitz-Rode, T.; Steinseifer, U. Evaluation of Hydraulic Radial Forces on the Impeller by the Volute in a Centrifugal Rotary Blood Pump. *Artif. Organs* **2011**, *35*, 818–825. [CrossRef] [PubMed]
11. Alemi, H.; Ahmad Nourbakhsh, S.; Raisee, M.; Farhad Najafi, A. Development of new "multivolute casing" geometries for radial force reduction in centrifugal pumps. *Eng. Appl. Comput. Fluid Mech.* **2015**, *9*, 1–11. [CrossRef]
12. Benra, F.K.; Dohmen, H.J.; Schneider, O. Calculation of hydrodynamic forces and flow induced vibrations of centrifugal sewage water pumps. In Proceedings of the ASME/JSME 2003 4th Joint Fluids Summer Engineering Conference, Honolulu, HI, USA, 6–10 July 2003; American Society of Mechanical Engineers: Hoston, TX, USA, 2003; pp. 603–608.
13. Benra, F.K.; Dohmen, H.J.; Schneider, O. Measurement of flow induced rotor oscillations in a single-blade centrifugal pump. In Proceedings of the ASME/JSME 2004 Pressure Vessels and Piping Conference, San Diego, CA, USA, 25–29 July 2004; American Society of Mechanical Engineers: Hoston, TX, USA, 2004; pp. 167–174.
14. Benra, F.K. Experimental investigation of hydrodynamic forces for different configurations of single-blade centrifugal pumps. In Proceedings of the 11th International Symposium on Transport Phenomena and Dynamics of Rotating Machinery (ISROMAC-11), Honolulu, HI, USA, 26 February–2 March 2006.
15. Benra, F.K. Numerical and Experimental Investigation on the Flow Induced Oscillations of a Single-Blade Pump Impeller. *J. Fluids Eng.* **2006**, *128*, 783–793. [CrossRef]
16. Aoki, M. Instantaneous Interblade Pressure Distributions and Fluctuating Radial Thrust in a Single-blade Centrifugal Pump. *Bull. JSME* **2008**, *27*, 2413–2420. [CrossRef]
17. Nishi, Y.; Fujiwara, R.; Fukutomi, J. Design Method for Single-Blade Centrifugal Pump Impeller. *J. Fluid Sci. Technol.* **2009**, *4*, 786–800. [CrossRef]
18. Nishi, Y.; Fukutomi, J.; Fujiwara, R. Radial Thrust of Single-Blade Centrifugal Pump. *IJFMS* **2011**, *4*, 387–395. [CrossRef]
19. Nishi, Y.; Fukutomi, J. Effect of Blade Outlet Angle on Unsteady Hydrodynamic Force of Closed-Type Centrifugal Pump with Single Blade. *Int. J. Rotating Mach.* **2014**, *2014*, 1–16. [CrossRef]
20. Tan, M.; Ji, Y.; Liu, H. Effect of Blade Wrap Angle on Performance of a Single-Channel Pump. *Exp. Tech.* **2018**, *2018*, 481–490. [CrossRef]
21. Chen, J.; Shi, W.; Zhang, D. Influence of blade inlet angle on the performance of a single blade centrifugal pump. *Eng. Appl. Comput. Fluid Mech.* **2021**, *15*, 462–475.
22. Tan, L.; Shi, W.; Zhang, D.; Zhou, L.; Wang, C. Numerical and experimental investigations of pressure fluctuations in single-channel pumps. *Proc. Inst. Mech. Eng. Part A J. Power Energy* **2018**, *232*, 97–415. [CrossRef]
23. Jiang, Z.; Li, H.; Shi, G.; Liu, X. Flow Characteristics and Energy Loss within the Static Impeller of Multiphase Pump. *Processes* **2021**, *9*, 1025. [CrossRef]
24. Yang, Y.; Zhou, L.; Shi, W.; He, Z.; Han, Y.; Xiao, Y. Interstage difference of pressure pulsation in a three-stage electrical submersible pump. *J. Pet. Sci. Eng.* **2020**, *196*, 107653. [CrossRef]
25. Yang, Y.; Zhou, L.; Hang, J.; Du, D.; Shi, W.; He, Z. Energy characteristics and optimal design of diffuser meridian in an electrical submersible pump. *Renew. Energy* **2021**, *167*, 718–727. [CrossRef]
26. Peng, G.; Huang, X.; Zhou, L.; Zhou, G.; Zhou, H. Solid-liquid two-phase flow and wear analysis in a large-scale centrifugal slurry pump. *Eng. Fail. Anal.* **2020**, *114*, 104602. [CrossRef]
27. Li, X.; Jiang, Z.; Zhu, Z.; Si, Q.; Li, Y. Entropy generation analysis for the cavitating head-drop characteristic of a centrifugal pump. *Proc. Inst. Mech. Eng. Part C J. Mech. Eng. Sci.* **2018**, *232*, 4637–4646. [CrossRef]
28. Tan, L.; Shi, W.; Zhang, D.; Wang, C.; Zhou, L.; Mahmoud, E. Numerical and experimental investigations on the hydrodynamic radial force of single-channel pumps. *J. Mech. Sci. Technol.* **2018**, *32*, 4571–4581. [CrossRef]

applied sciences

MDPI

Article

Compact Model of a Screen under Fan-Induced Swirl Conditions Using a Porous Media Approach

Asier Bengoechea [1,2], Raúl Antón [2], Alejandro Rivas [2], Gorka S. Larraona [2] and Juan Carlos Ramos [2,*]

1 Sarralle Steel Melting Plant, Engineering Department, Orendaundi Kalea 7, 20730 Azpeitia, Spain; abengoetxea@sarralle.com

2 Thermal and Fluids Engineering Division, Mechanical Engineering and Materials Department, TECNUN–School of Engineering, University of Navarra, Paseo de Manuel de Lardizábal 13, 20018 San Sebastián, Spain; ranton@tecnun.es (R.A.); arivas@tecnun.es (A.R.); gsanchez@tecnun.es (G.S.L.)

* Correspondence: jcramos@tecnun.es

Abstract: A perforated plate in an electronic device is typically placed downstream of an axial fan (push cooling) in order to avoid electromagnetic interferences. Because of the swirling component in the flow approaching the screen, determining how the screen affects the flow pattern downstream of the screen is a challenge. It is important to understand this interaction, as the correct location of the electronic components will depend on the flow pattern (the components that dissipate more heat will be located where the maximum magnitude of the velocity is located). This work aims to present an approach of the flow pattern via a compact model based on three directional pressure loss coefficients. The values for the pressure loss coefficients are obtained through different correlations depending on the flow and geometric characteristics for the case that is being modeled. These correlations are obtained through an iterative process that compares different flow patterns obtained through different modeling strategies: the compact one that is presented in this paper and another detailed one, which was validated in previous works. Results show that if this compact model is used, an approximation of the flow pattern could be obtained with a huge decrease in the amount of time invested.

Keywords: flow pattern; swirl flow; computational fluid dynamics (CFD); perforated plate; pressure loss coefficients; compact model

Citation: Bengoechea, A.; Antón, R.; Rivas, A.; Larraona, G.S.; Ramos, J.C. Compact Model of a Screen under Fan-Induced Swirl Conditions Using a Porous Media Approach. *Appl. Sci.* **2021**, *11*, 1999. https://doi.org/10.3390/app11051999

Academic Editor: Jesús María Blanco

Received: 22 January 2021
Accepted: 20 February 2021
Published: 24 February 2021

1. Introduction

Thermal management in electronic devices has become an important issue because of the huge increase in heat generation. It is normal to place an electronic device in an electronic cabinet or rack that has several sub-racks inside it, upon which the printed circuit boards (PCBs) are placed. The heat generated by the PCBs must be removed from each of these levels via an appropriate cooling strategy. There are several cooling strategies (liquid cooling [1,2], two-phase cooling [3] and spray cooling [4,5], among others), but the most commonly employed technique is air cooling using an axial fan to produce air movement for reasons of safety and availability. Problems arise as the axial fan is usually placed together with an electromagnetic compatibility (EMC) screen, which mainly plays the role of reducing electromagnetic interferences in the device but it also keeps dust from entering.

The configuration of electronic devices has changed a lot over the years. Several years ago, the axial fan typically worked in a push or pull configuration. However, although some devices continue to use these configurations, nowadays, a pull–push configuration has become more common because of the huge increase in heat generation. The push–pull configuration is depicted in Figure 1.

For each axial fan in the figure, there is an associated difficulty. The rightmost fan in Figure 1, which sucks air out of the system, might be hydraulically affected if the EMC screen is placed too close to the fan (although Figure 1 presents downstream placement, it is possible to place the screen either upstream or downstream from the fan). The hydraulic influence means that the characteristic curve of a fan (static pressure versus volumetric flow rate) is reduced. This effect in electronic cooling has been studied by several authors, such as Hill [6], Grimes et al. [7], Lin and Chou [8] and Swim [9], and it has also been studied for automotive cooling purposes by Baniasadi et al. [10]. Antón et al. [11] study the effect of various parameters, and they also give the correlations for predicting the effect of the main parameters (porosity of the screen and distance between fan and screen) for the points of the characteristic curve with the highest static pressure (Q = 0) and volumetric flow ($\Delta P_s = 0$). In a similar way, the characterization of the pressure drop coefficients of perforated plates in ducts by means of numerical simulations of complete detailed models was addressed by Tanner et al. [12], who obtained correlations of the pressure drop as a function of the Reynolds number and using as parameters the porosity and the thickness of the plates, or Miguel [13], where the influence of the porosity and the Reynolds numbers on the pressure loss factors was analyzed.

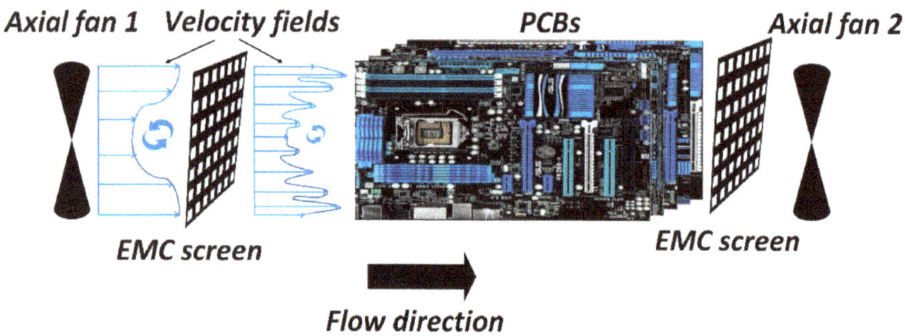

Figure 1. Scheme of a pull–push cooling configuration.

It is extremely important to know the velocity field downstream from the first EMC screen (see Figure 1), since the cooling capacity for a given electronic component is related to the local velocity magnitude near that component [14]. The pressure drop through the screen is also of interest in order to select a proper axial fan. However, due to the complexity of the approaching flow (an annular profile together with a high rotational component, as depicted in Figure 1), there is a high degree of uncertainty when predicting the downstream velocity field and pressure drop through the screen. It should be noted that the EMC screen modifies the relationship between the axial and tangential components of the flow, which makes predicting the velocity field at the outlet of the screen a challenge [15].

Bengoechea et al. [16] analyzed this issue, investigating the flow change through an EMC screen when the approaching flow is a fan-induced swirling flow. They analyzed two very distinct cases: from the mean flow and screen porosity point of view, among other parameters. They measured the flow pattern after the screen using the particle image velocimetry (PIV) technique, and they then used the experimental values for a computational fluid dynamics (CFD) benchmark. The benchmark conclusion was that for this complex flow, the turbulence model that best predicts the flow was the Reynolds stress transport model (RSTM). The domain of the CFD model was a detailed model, a model in which all the holes or pores of the EMC screen were created and meshed. At the same time, they defined three flow magnitudes that together are able to properly define a velocity field with an annular shape. Once the RSTM was determined to be the correct choice, in Bengoechea et al. [17], a parametric study was presented in order to analyze the influences that different parameters have on the variation of the velocity field that exists between the upstream and downstream of a screen and in the total pressure drop. That parametric study was based on 81 detailed (in which the geometry of the screens with their holes was defined) CFD simulations defined by a Design of Experiments (DoE) approach. Additionally, the study presented, in conjunction with a reduced model for predicting the total pressure drop, reduced models based on DoE techniques [18] for predicting the variation in the flow magnitudes that Bengoechea et al. [16] defined in order to set the velocity field. These reduced models could be of great help to thermal designers by allowing them to have a rough idea of how the EMC screen will affect the velocity field and the total pressure drop. Nevertheless, thermal designers will not have an approximation of the velocity field, and so in order to attain that, they will have to run simulations with detailed geometries, such as those used in Bengoechea et al. [16,17]; not only does this require high computational demands, calculating the geometry and the mesh is also a highly time-consuming task. Because product launching times are being progressively reduced, working with compact models of the components in an electronic system is becoming a key modeling approach. The EMC screen is among the components that must be modeled in a compact way, and this is the aim of the present work.

The compact modeling approach is based on directional pressure loss coefficients that define a porous media domain that is very easy to create and mesh. However, the fact that the holes of the screen are not created in a compact model has some drawbacks [19], a major one being that the information about local flow phenomena in very close proximity to a real screen is lost (the small jets produced after the screen, see Figure 1). However, the drawbacks are small in comparison to the benefits, which include, among other things, the reduction in computational demands. Thermal designers may model the screen as a planar resistance, using only one pressure loss coefficient from handbooks like Idelchick's [20]. Kordyban [21] was the first to analyze the need to model a screen as a volumetric resistance, which meant employing more than one coefficient. The results of modeling the screen as a planar and as a volumetric resistance were compared, and it was concluded that if the flow approaching the screen has any tangential component, a planar resistance is not the correct choice. This is because a planar resistance makes the flow more parallel to the screen due to the fact that the unique pressure loss coefficient only affects the axial component of the velocity. The same need to employ more than one coefficient was also shown by Antón et al. [22–25]. In Nevelsteen et al. [26], a volumetric model of a screen was experimentally validated, and a satisfactory fitting was attained. Nevertheless, only one experimental case was studied, and therefore, it was not possible to obtain the pressure loss coefficients as a function of the experimental parameters.

This article uses an iterative process (taking as a reference the results of the 81 detailed numerical cases from Bengoechea et al. [17]) to introduce correlations for the pressure loss coefficients per unit length of a compact model of a screen as a function of different geometric and operational parameters.

The use of compact models to characterize the performance of complex systems in order to make the simulation and design tasks easier and faster is widely extended in different engineering applications: the electric field generated in tunnel field-effect transistors, Najam et al. [27]; the circuit models of vertical-cavity surface-emitting lasers, Li et al. [28]; the dynamic performance of latent heat thermal storage units, Colangelo et al. [29].

The compact model developed and presented here will aid thermal designers, as it will reduce the time needed to create the model, especially when simulating. Specifically, this article presents a powerful tool that makes it possible to obtain a fast and reliable approximation of the velocity field downstream of a screen.

2. Geometric and Operational Parameters for the Numerical Runs

The process for obtaining the correlations for the pressure loss coefficients per unit length, as will be explained later, employs the results from the 81 detailed numerical runs presented in [17]. Therefore, the geometric and operational parameters in the numerical runs for this work must be the same as those employed in that work. The parameters and their range are presented in Table 1. "D" is the size of the channel, which corresponds to the diameter of the shroud of the fan being analyzed (see Figure 2). "D_h/D" is the ratio between the hub of the fan and the shroud diameter (see Figure 2). "S" is the swirl number as defined by Beér and Chigier [30] and given in Equation (1). In this definition, L_c is a characteristic length assumed to be half of the simulated fan diameter in this work; "r" is the radial coordinate; and "V_a" and "V_t" are the local axial and tangential velocities, respectively. "a" is the distance between the outlet of the fan (in this work, the outlet of the fan is the domain inlet) and the inlet of the screen (see Figure 2). "ε" is the porosity of the screen, defined as the ratio between the open area and the total area of the cross section ($D \times D$). The next parameter, "ξ", is the ratio between the total lateral perimeter of the pores and the area of the cross-section. "t" is the thickness of the screen (see Figure 2), and "V" is the mean axial velocity in the channel. The 81 numerical cases are shown in Table 2; the "-", "0" and "+" level values used for each parameter are defined in Table 1.

$$S = \frac{\iint_A \rho \cdot V_a \cdot V_t \cdot r \cdot dA}{L_c \cdot \iint_A \rho \cdot V_a^2 \cdot dA} \quad (1)$$

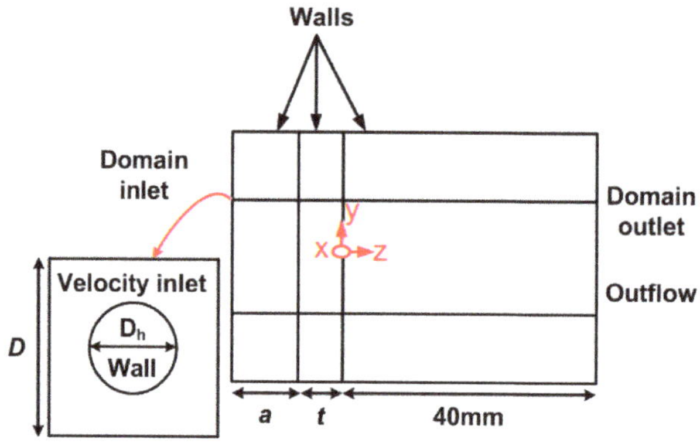

(a)

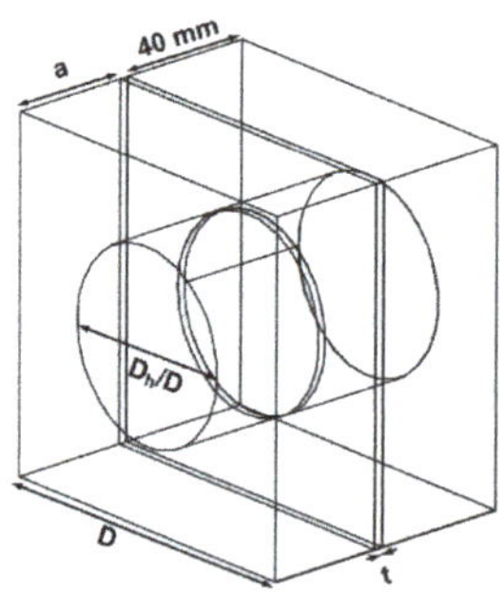

(b)

Figure 2. (**a**) Generic scheme of the domain and boundary conditions used for the simulations; (**b**) 3D representation of the flow domain.

Table 1. Operational and geometric parameters included in the study and their range.

Parameters	D (mm)	D_h/D (-)	S (-)	a (mm)	ε (%)	ξ (1/mm)	t (mm)	V (m/s)
Minimum value (-)	40	0.25	0.4	5	30	0.25	0.5	1
Middle value (0)	70	0.4	0.7	15	52.5	0.31	1.5	1.75
Maximum value (+)	100	0.55	1	25	75	0.37	2.5	2.5

Table 2. Numerical runs established by the Design of Experiments (DoE) to be simulated in computational fluid dynamics (CFD).

Exp.	D	D_h/D	S	V	a	t	ε	ξ	Exp.	D	D_h/D	S	V	a	t	ε	ξ
1	-	+	-	-	+	+	-	-	**42**	-	+	+	-	+	-	+	+
2	-	-	+	+	-	-	+	+	**43**	+	+	-	+	-	-	-	+
3	0	0	+	0	0	0	0	0	**44**	-	-	-	-	-	+	+	-
4	0	0	0	0	-	0	0	0	**45**	-	+	+	+	+	+	-	-
5	+	+	+	+	+	-	+	-	**46**	+	+	-	-	-	-	+	+
6	+	+	-	+	-	+	-	-	**47**	-	-	-	+	-	+	-	-
7	0	0	0		0	0	0	0	**48**	+	-	+	+	-	+	-	+
8	+	-	+	+	-	-	-	-	**49**	-	+	+	-	+	+	+	-
9	0	0	0	0	0	-	0	0	**50**	-	+	+	+	-	+	-	+
10	+	+	+	-	+	+	-	+	**51**	0	+	0	0	0	0	0	0
11	-	0	0	0	0	0	0	0	**52**	-	+	-	+	+	-	+	+
12	0	0	-	0	0	0	0	0	**53**	-	-	-	-	+	+	+	+
13	-	+	+	+	+	-	-	+	**54**	+	-	-	+	+	+	+	-
14	+	-	+	-	-	+	+	+	**55**	0	0	0	0	+	0	0	0
15	+	-	-	-	+	+	-	-	**56**	+	+	-	-	+	+	+	+
16	-	-	-	-	+	-	+	-	**57**	0	0	0	0	0	0	0	0
17	0	0	0	0	0	0	-	0	**58**	-	+	+	-	-	+	+	+
18	+	-	-	-	+	-	-	+	**59**	-	+	-	+	-	+	+	+
19	-	+	+	-	-	-	+	-	**60**	+	+	+	+	+	+	+	+
20	0	0	0	0	0	0	+	0	**61**	-	-	+	-	-	+	-	-
21	-	+	-	-	-	+	-	+	**62**	+	+	+	-	-	+	-	-
22	+	+	-	-	-	+	+	-	**63**	-	-	+	+	-	+	+	-
23	+	-	-	+	+	-	+	+	**64**	0	0	0	0	0	0	0	+
24	+	-	-	-	-	-	-	-	**65**	-	-	+	-	-	-	-	+
25	+	-	-	-	-	+	-	+	**66**	+	-	-	+	-	+	+	+
26	-	+	-	-	-	-	-	-	**67**	+	+	+	+	-	+	+	-
27	0	0	0	0	0	0	0	-	**68**	0	-	0	0	0	0	0	0
28	+	+	-	+	+	-	-	-	**69**	+	+	-	+	+	+	-	+
29	+	-	+	+	+	+	-	-	**70**	+	-	+	+	+	-	-	+
30	+	-	+	-	+	-	+	+	**71**	-	-	+	-	+	+	-	+
31	-	+	-	+	+	+	+	-	**72**	+	+	+	-	-	-	-	+
32	+	+	+	-	+	-	-	-	**73**	-	-	-	+	-	-	-	+
33	0	0	0	+	0	0	0	0	**74**	-	+	-	-	+	-	-	+
34	-	-	+	-	+	-	-	-	**75**	-	-	+	+	+	+	+	+
35	+	0	0	0	0	0	0	0	**76**	+	+	-	-	+	-	+	-
36	-	-	-	+	+	-	-	-	**77**	-	-	-	-	-	-	+	+
37	+	-	+	-	+	+	+	-	**78**	-	+	+	+	-	-	-	-
38	-	+	-	+	-	-	+	-	**79**	+	-	-	+	-	-	+	-
39	-	-	+	+	+	-	+	-	**80**	+	-	+	-	-	-	+	-
40	-	-	-	+	+	+	-	+	**81**	0	0	0	0	0	+	0	0
41	+	+	+	+	-	-	+	+									

3. Numerical Model

3.1. Flow Domain

In contrast to the 3D perforated plates with squared holes modeled in Bengoechea et al. [17], the simple geometry depicted in Figure 2 is employed as a flow domain for each of the compact models for the 81 cases from Table 2. The novelty of this work is modeling the screen via two different porous media volumes (the inner cylindrical volume and the outer one, as shown in Figure 2). It is to these volumes that the pressure loss coefficients (the objective of this work) are applied. As the geometry of the pores does not have to be modeled, the time needed to create the geometry is substantially reduced. In order to build a conformal mesh between the screen and the downstream and upstream volumes, these volumes are also divided into two volumes in such a way that the same mesh strategy is used throughout the entire volume. The length of the volumes upstream of the screen have the value of parameter "*a*" (see Figure 2) for each case. The annulus shape that a flow has at the outlet of a fan (see Figure 1), which is needed at the inlet of the domain, is obtained following Bengoechea et al. [17] by placing a circular wall whose diameter is "D_h" at the inlet cross-section. Consequently, the inlet velocity conditions are imposed on the area around it. In order to make it possible to compare the results of the detailed models from Bengoechea et al. [17] with the compact models from the present work, the channel length after the screen must be the same as that used in the first study. To that end, a value of 40 mm was fixed.

3.2. Mathematical Modeling

3.2.1. Turbulence Modeling

In previous experimental measurements carried out by the authors in a wind tunnel [16], it was confirmed that the flow that is being modeled is fully turbulent throughout the domain under consideration. The Reynolds-averaged Navier-Stokes (RANS) approach was adopted for numerically simulating the flow, and thus the mean flow was calculated by solving Equations (2) and (3):

$$\frac{\partial U_i}{\partial x_i} = 0 \tag{2}$$

$$\frac{\partial}{\partial x_j}(\rho U_i U_j) = -\frac{\partial P}{\partial x_i} + \frac{\partial}{\partial x_j}\left[\mu\left(\frac{\partial U_i}{\partial x_j} + \frac{\partial U_j}{\partial x_i}\right) - \rho\overline{u'_i u'_j}\right] + S_i \tag{3}$$

In order to solve the Reynolds stresses ($\rho\overline{u'_i u'_j}$) and subsequently close the system of Equations (2) and (3), the standard k-ε model was applied [31]. Although the detailed models in Bengoechea et al. [16,17] used the RSTM as a turbulence model, the standard k-ε model could be used in this work as the values of the pressure loss coefficients are those that best fit the results obtained in the detailed simulations (those that model the holes). This may be seen as a limitation, but it was done on purpose. The k-ε model was chosen due to its robustness and lower cost from a computational point of view, and, therefore, it is of more interest for the thermal designers. Since the procedure consists of an iterative process that serves as a calibration of the coefficients for each of the 81 cases, it can be tuned in such a way that takes into consideration not only the porosity physics but also the difference between the two turbulence models.

Within the standard k-ε model, the Reynolds stresses are calculated by means of the Boussinesq hypothesis:

$$\rho\overline{u'_i u'_j} = -\mu_t\left(\frac{\partial U_i}{\partial x_j} + \frac{\partial U_j}{\partial x_i}\right) + \frac{2}{3}\rho k \delta_{ij} \tag{4}$$

So that the Reynolds stress tensor is aligned with the mean strain-rate tensor (S_{ij}). The standard k-ε model consists of two transport equations for both the turbulent kinetic energy (k) and its dissipation rate (ε), which can be expressed in the following compact way:

$$\frac{\partial}{\partial x_i}(\rho U_i k) = \frac{\partial}{\partial x_j}\left(\Gamma_k \frac{\partial k}{\partial x_j}\right) + P_k - D_k \tag{5}$$

$$\frac{\partial}{\partial x_i}(\rho U_i \varepsilon) = \frac{\partial}{\partial x_j}\left(\Gamma_\varepsilon \frac{\partial \varepsilon}{\partial x_j}\right) + P_\varepsilon - D_\varepsilon \tag{6}$$

The terms used in both equations are defined in Table 3. The turbulent viscosity is calculated as

$$\mu_t = \rho C_\mu \frac{k^2}{\varepsilon} \tag{7}$$

with a value of 0.09 for C_μ.

Standard wall functions (SWF), which are based on the work by Launder and Spalding [32], were used for the near-wall treatment of the turbulence model used to predict flow characteristics very close to solid walls. This wall treatment does not require a fine meshing close to the wall, which means there is a significant reduction in the amount of elements in the model and, therefore, a reduction in computational time.

Table 3. Terms and coefficients for the standard k-ε model.

Standard k-ε	
Γ_k	$\mu + \frac{\mu_t}{\sigma_k}$
Γ_ε	$\mu + \frac{\mu_t}{\sigma_\varepsilon}$
P_k	$\mu_t S^2$
D_k	$\rho\varepsilon$
P_ε	$C_{1\varepsilon}\frac{\varepsilon}{k}P_k$
D_ε	$C_{2\varepsilon}\rho\frac{\varepsilon^2}{k}$
Coeff.	$\sigma_k = 1.0; \sigma_\varepsilon = 1.3$ $C_{1\varepsilon} = 1.44$ $C_{2\varepsilon} = 1.92$

3.2.2. Porous Media Modeling

Both volumes corresponding to the screen (the inner and the outer volumes with "t" thickness, as shown in Figure 2) are assumed as porous media volumes by using directional pressure loss coefficients (φ_x, φ_y and φ_z) per unit length [33]. The same coefficients ($\varphi_x = \varphi_y = \varphi_p$) were used for the "$x$" and "$y$" directions, and this value is the same for the inner and outer volumes ($\varphi_{p\ int} = \varphi_{p\ ext} = \varphi_p$), resulting in a model with greater simplicity. Therefore, the objective of the work is to obtain the pressure loss coefficients "φ_p", "$\varphi_{z\ int}$" and "$\varphi_{z\ ext}$" per unit length that define a compact model.

The modeling of a screen through the porous media approximation employs a volumetric hydraulic resistance defined by the previous directional pressure loss coefficients per unit length. In the following expression, the generic strain that must be fixed for both the inner and outer porous media volumes in each of the 81 cases from Table 2 is given:

$$\varphi_{ij} = \begin{pmatrix} \varphi_p & 0 & 0 \\ 0 & \varphi_p & 0 \\ 0 & 0 & \varphi_z \end{pmatrix} \tag{8}$$

This porous media approach works as a momentum sink in the momentum equation. This sink usually has two different components: the first one is related to the viscous losses (proportional to the velocity), and the second one is due to inertial losses (proportional to the squared velocity). For this study, with the EMC screens, viscous loses are neglected as the inertial losses are far higher. Therefore, the momentum sink might be defined as

$$S_i = -\left(\varphi_{ij}\cdot\frac{1}{2}\cdot\rho\cdot v_{mag}\cdot v_j\right) \tag{9}$$

where v_{mag} corresponds to the magnitude of the velocity and v_j to the component of the velocity in "y" direction.

Due to the fact that the interior walls of the pores are not modeled in the porous media, a special treatment for the velocities is needed. For this reason, by ensuring the continuity of the vectors through the porous media, a superficial velocity based on the volumetric flow is employed. Therefore, the physics inside the compact model are not the same as those that occur in reality or in the detailed models. In fact, compact models are not able to correctly predict the jets produced by the pores (see the velocity field at the outlet of the screen in Figure 1). This is the reason why the compact models introduced in this work will be reliable at a certain distance from the screen once the flow has been homogenized and the jets have disappeared.

3.3. Boundary Conditions

The scheme of the employed boundary conditions for the compact modeling is shown in Figure 2a, together with the established coordinate system (placed at the center of the cross-section at the outlet of the screen, $z = 0$). The tangential velocity is defined as having an origin around the $x = 0$ and $y = 0$ points.

The non-slip condition was fixed for the walls, and an outflow condition was fixed at the outlet cross-section of the domain. For the velocity at the inlet, different constant values of axial and tangential velocities were imposed for each of the 81 cases from Table 2. The constant values of axial and tangential velocity for each case are such that the mean axial velocity (V) and the swirl number (S) conditions at the inlet coincide with the values shown in Table 2, and they are the same values employed by Bengoechea et al. [17].

The turbulence intensity also must be fixed as conducted by Bengoechea et al. [17]. Thus, at the inlet for all cases, the turbulence was fixed through the hydraulic diameter (D_h) of the section (D, as it is a square section), and a constant turbulence intensity of 50%. In that work, it was explained that changes in the values of turbulence intensity at the inlet did not produce important changes in the flow pattern downstream of a screen and, therefore, this inlet turbulence modeling strategy was validated.

4. Numerical Simulation of the Model

4.1. Geometry Meshing

Since the geometry in this compact model is very simple compared to the detailed model employed in Bengoechea et al. [16,17], the time required to generate the geometry and mesh it for each of the 81 cases from Table 2 did not exceed 5 min.

Before running any simulations, a mesh sensitivity study was carried out in order to set a correct mesh size for each of the model sizes: 40, 70 and 100 mm (see values for D in Table 1). Therefore, cases 1 and 2 from Table 2 were simulated with random values for the pressure loss coefficients and different interval sizes (IS) for the mesh (0.2, 0.3, 0.4, 0.5 and 0.6 mm). It must be highlighted that the thickness parameter (t) for cases 1 and 2 are at different levels, and therefore, it was also necessary to check whether screen thickness might have any influence on the mesh size requirements. The difference between the model using an IS of 0.2 and 0.6 was less than 6% in the peak for axial or tangential velocities at the centerline ($x = 0$) of a cross-section located 20 mm downstream from the screen. Additionally, it was necessary to verify that a low thickness did not demand a higher density for the mesh. Therefore, as the main objective of the compact model is to reduce the time required, an IS of 0.6 mm was chosen as the mesh resolution for all the models with a 40 mm value for parameter D.

Once this IS was established for the small size in all cases, the same study was carried out for case 6 from Table 2, where the model size was fixed to its higher value (100 mm). The mesh sizes simulated for this case were 0.4, 0.8, 1.0 and 1.5 mm. Similar to cases 1 and 2, because the variation was not greater than 6.7% for the peak values (in $x = 0$ for the 20 mm downstream cross-section), an IS of 1.5 mm was selected as the correct choice.

Finally, for the middle value of the diameter (70 mm), the growth rate was respected. For the low value of D (40 mm) an IS of 0.6 mm was used, and for the high value (100 mm), an IS of 1.5 mm was used, and thus the growth rate was 2.5 (100 mm/40 mm = 1.5/0.6 mm). Similarly, for the middle value of D (70 mm), an IS of 1.05 mm was employed (70/40 mm = 1.05/0.6 mm). A sample of the mesh used is presented in Figure 3.

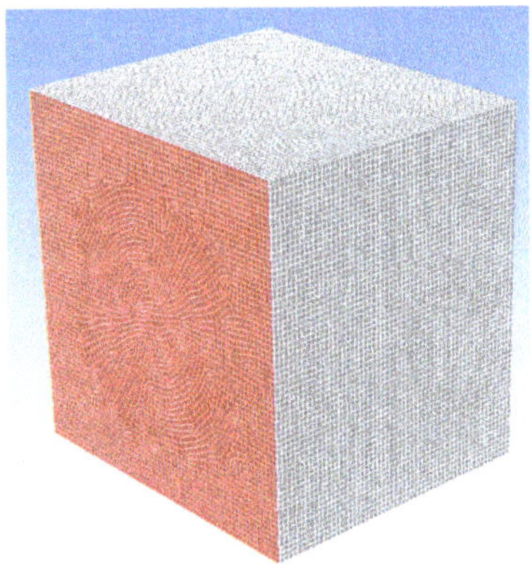

Figure 3. Representation of the mesh for one of the cases simulated of the compact model.

4.2. Numerical Procedure

The finite volume method implemented in Ansys Fluent® 13.0 [33] was used as the numerical discretization method. A segregated scheme using a second-order centered scheme for the diffusion terms was used when solving the equations of the mathematical models. A second-order upwind scheme was employed to discretize the convection terns of the RANS equations and the transport equations for the turbulence quantities k and epsilon. The SIMPLE pressure–velocity–coupling algorithm was employed to solve the continuity equation by using a momentum interpolation method to calculate velocities at cell faces.

4.3. Convergence Criterion and Computational Time

In order to set an adequate level of convergence, the guidelines proposed by the European Research Community on Flow, Turbulence and Combustion [34] were followed. This way, when stating that a case was converged, the scaled residuals of all the variables were always less than 1.1×10^{-5}. Different flow magnitudes were monitored in a cross-section located 20 mm downstream from the screen in each case, and it was verified that there was no change in the 4th significant digit when stopping the simulations.

To perform the simulations, an HP Proliant DL 585 G7 server (DL 585 G7, HP, Leon, Spain, 2013) that had 32 processors at 2.8 GHz, 64 GB of shared RAM memory and ran the Linux operating system was used. The communication library used was Platform MPI, and the required CPU time for the simulation was below one minute. It must be highlighted that for the simulations by Bengoechea et al. [17], the computational time with the same server oscillated between 66 and 360 h. Moreover, even using a not very powerful computer, no more than 10 min (less than 0.25% of the detailed model computation time) is needed in order to reach a correct convergence.

5. Iterative Process to Obtain the Pressure Loss Coefficients per Unit Length

The objective of the iterative process is to obtain the values for the pressure loss coefficients for each of the 81 cases introduced in Table 2. As such, the same iterative process is applied to each of the 81 cases.

The process for each of the cases starts with 5 initial simulations with randomized and different values for "φ_p", "$\varphi_{z\ int}$" and "$\varphi_{z\ ext}$". This way, the response of the model (which could be described as a response surface of the coefficients) to coefficient values changes is captured. From each of the 5 simulations, values are obtained for the area weighted average of the tangential velocity "$\overline{v_t}$", the swirl number "S" and the minimum and the maximum axial velocities "V_{min}" and "V_{max}" from the outlet cross-section (z = 40 mm in Figure 2). These magnitudes are the flow magnitudes that Bengoechea et al. [16] defined in order to correctly define a flow pattern with an annular shape. Therefore, it might be said that the values of the flow magnitudes in each of the 5 simulations correctly define the different flow patterns (as a result of the different pressure loss coefficients) at the outlet cross-section of the model. The next step is to find a function that represents the response surface model. That function can be described by obtaining the values *A*, *B*, *C* and *D* in Equation (10). This equation represents the relationship between the pressure loss coefficients ($\varphi_{p\ i}$, $\varphi_{z\ int\ i}$ and $\varphi_{z\ ext\ i}$) used for the five initial simulations and the result obtained for ϕ_i in the five cases, where ϕ_i is one of the flow magnitudes ($\overline{v_t}$, S, V_{min} and V_{max}):

$$\begin{pmatrix} \phi_1 \\ \phi_2 \\ \phi_3 \\ \phi_4 \\ \phi_5 \end{pmatrix} - (A\ \ B\ \ C) \begin{pmatrix} \varphi_{p\ 1} & \varphi_{p\ 2} & \varphi_{p\ 3} & \varphi_{p\ 4} & \varphi_{p\ 5} \\ \varphi_{z\ int\ 1} & \varphi_{z\ int\ 2} & \varphi_{z\ int\ 3} & \varphi_{z\ int\ 4} & \varphi_{z\ int\ 5} \\ \varphi_{z\ ext\ 1} & \varphi_{z\ ext\ 2} & \varphi_{z\ ext\ 3} & \varphi_{z\ ext\ 4} & \varphi_{z\ ext\ 5} \end{pmatrix} - D = 0 \quad (10)$$

Thus, for each of the flow magnitudes, a group of values for *A*, *B*, *C* and *D* is obtained. With these values, a new step is carried out to search for the values for φ_p, $\varphi_{z\ int}$ and $\varphi_{z\ ext}$ that minimize Equation (11) according to the response surface (*A*, *B*, *C* and *D*) that has been obtained for each of the flow magnitudes. In other words, a combination of pressure loss coefficients that create a flow pattern and pressure drop like that obtained in the detailed model is sought. Equation (11) consists of a summary of the differences (Equations (12)–(15)) obtained when comparing the flow magnitudes obtained with the detailed model from Bengoechea et al. [16] and those obtained with the compact model whose coefficients are being determined. The solution that is obtained ($\varphi_{p\ i}$, $\varphi_{z\ int\ i}$ and $\varphi_{z\ ext\ i}$) is the one that minimizes Equation (11) according to the response surface (*A*, *B*, *C* and *D*) obtained in the previous step.

$$F = \frac{Abs(\Delta \overline{v_t}) + Abs(\Delta S) + Abs(\Delta v_{max}) + Abs(\Delta v_{min})}{4} \quad (11)$$

$$\Delta \overline{v_t} = \left(1 - \frac{\overline{v_t}_{compact}}{\overline{v_t}_{detailed}}\right) \times 100 \quad (12)$$

$$\Delta S = \left(1 - \frac{S_{compact}}{S_{detailed}}\right) \times 100 \quad (13)$$

$$\Delta v_{max} = \left(1 - \frac{v_{max\ compact}}{v_{max\ detailed}}\right) \times 100 \quad (14)$$

$$\Delta v_{min} = \left(\frac{Abs(v_{min\ compact} - v_{min\ detailed})}{V}\right) \times 100 \quad (15)$$

With the new values obtained for $\varphi_{p\ i}$, $\varphi_{z\ int\ i}$ and $\varphi_{z\ ext\ i}$, a new simulation is run using these values for the pressure loss coefficients. Thus, the values for the flow magnitudes in this new simulation are obtained, and a new row is added to the system in (10), and then, the response surfaces are recalculated (*A*, *B*, *C* and *D*) in such a way that a higher accuracy is achieved based on the new additional case. Subsequently, new values for the coefficients are determined, and then the response surfaces are recalculated, thus beginning the iterative process. The break point of the iterations is when the variation in the values for the pressure loss coefficients between two iterations is negligible.

6. Correlations for the Pressure Drop Coefficients per Unit Length and Reliability of the Compact Model

6.1. Different Trends for the Behavior of the Value for $\varphi_{z\ ext}$

The iterative process was completed for the 81 cases in Table 2. Once the values for the coefficients per unit length were obtained for all cases, it was possible to distinguish three different trends in the values of the pressure loss coefficients among the 81 cases. The first trend includes only two cases, for which a correct adjustment between the compact and the detailed models was obtained using a value for $\varphi_{z\ ext}$ that was higher than 0 and lower than $\varphi_{z\ int}$. An example of this is presented in Figure 4, where the results for case 40 ($\varphi_{z\ ext} = 3533$ and $\varphi_{z\ int} = 100,000$) from Table 2 are shown. As the figure illustrates, the approach of the velocity field that was obtained presents a satisfactory adjustment with that obtained by the detailed modeling from Bengoechea et al. [17].

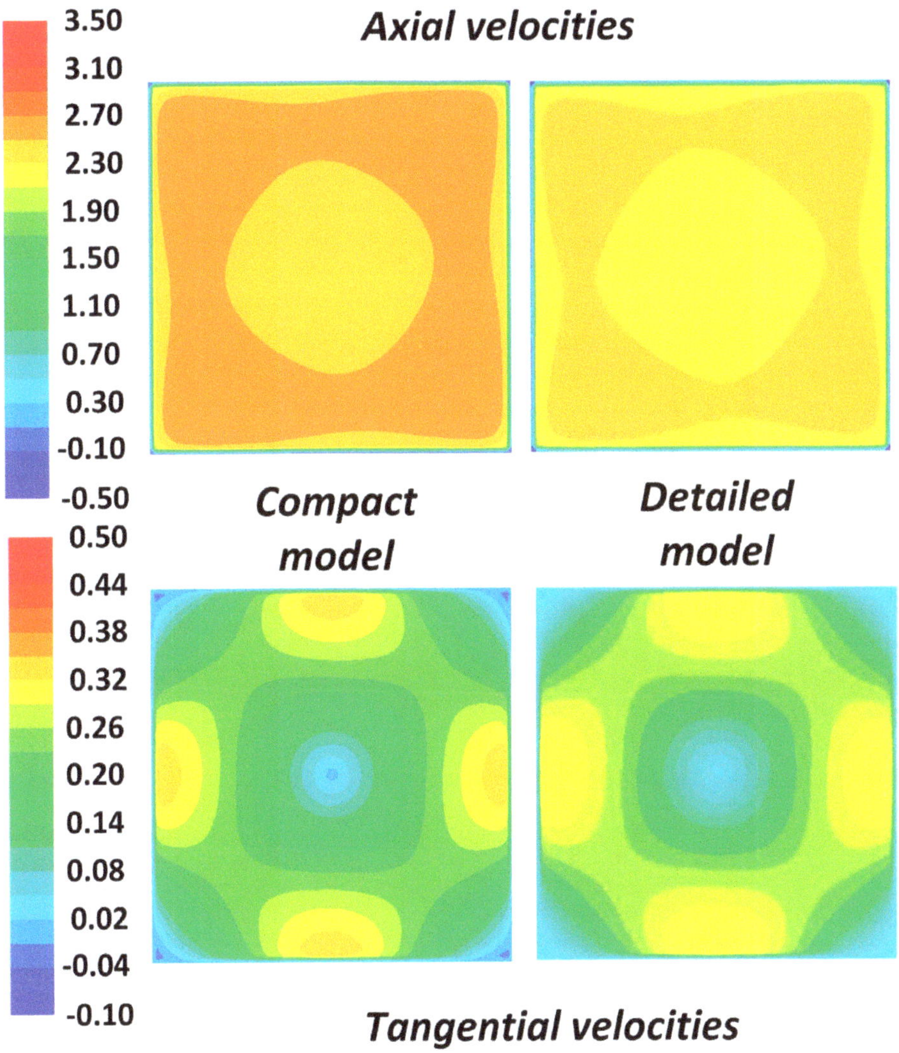

Figure 4. Axial and tangential velocity patterns (m/s) for the outlet cross-section obtained for case 40 in Table 2 via compact modeling and detailed modeling.

In the second trend, negative values for $\varphi_{z\ ext}$ are needed to achieve a correct adjustment. Using a negative value for this coefficient does not make any physical sense, as it means a momentum is generated instead of a sink, which is what it should be. However, there is some similarity with what actually happens in the detailed models, as they show that the flow by the hub zone is deviated towards the annulus side when crossing a screen in a certain way (as can be seen in Figure 5). In other words, the annulus side receives momentum from the hub side. The only way to generate this effect for compact modeling through porous media is by using a negative value for $\varphi_{z\ ext}$.

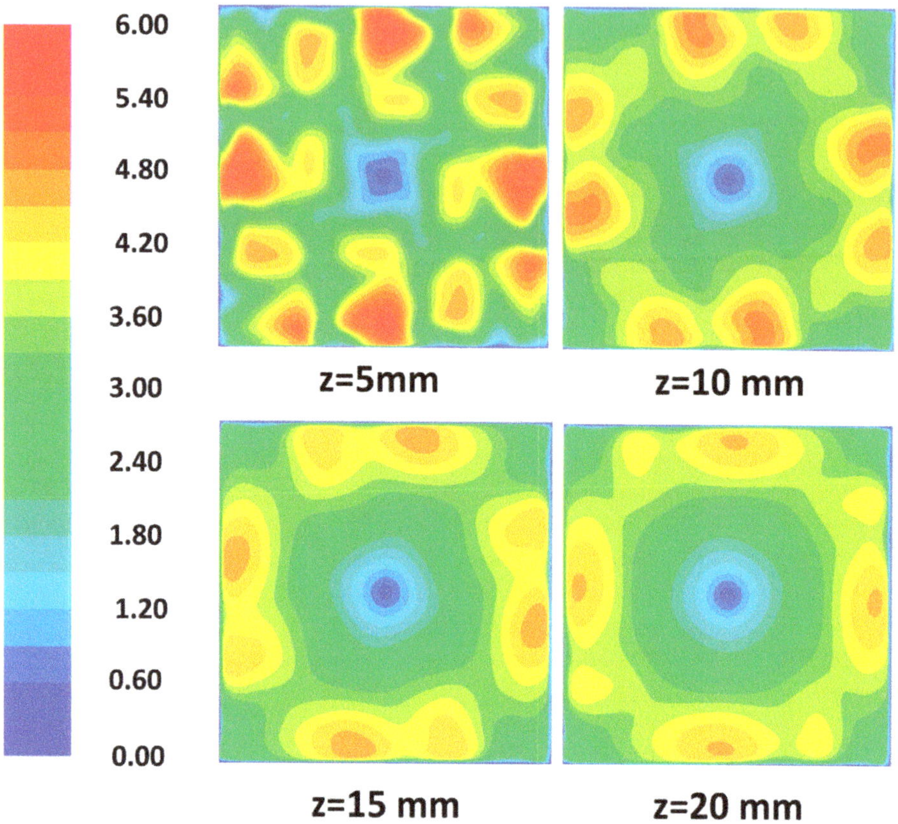

Figure 5. Velocity magnitudes contour plots (m/s) at different distances from the outlet of the screen for case 2 in Table 2.

The rest of the cases belong to the third trend, which includes all those cases where a very good adjustment is not achieved for the minimum and maximum axial velocity. As the $\varphi_{z\ ext}$ coefficient is reduced further, a correct adjustment is achieved for the maximum velocity but not for the minimum one, for which the value of $\varphi_{z\ ext}$ should be even smaller. An example of this trend is shown in Figure 6, which shows the second case in Table 2. The adjustment presented between both models in the figure indicates that the approach is good even for the difference in the minimum velocity value. The maximum axial velocity zone and the tangential velocity field are well approximated.

It must be highlighted that the values for the coefficients were obtained by comparing the flow magnitudes at the outlet cross-sections of the domain for all cases. However, the approaches of the flow pattern obtained by this compact modeling were verified as being reliable once the jets produced by the holes disappeared.

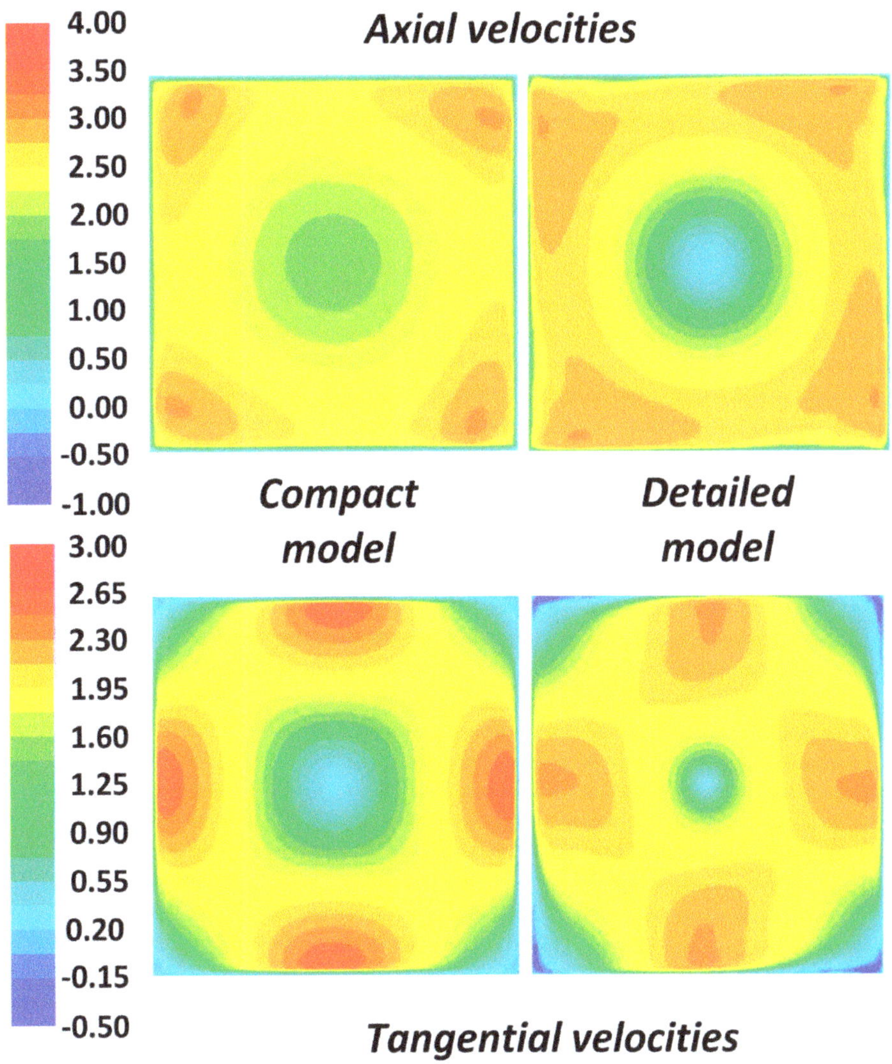

Figure 6. Axial and tangential velocity patterns (m/s) for the outlet cross-section obtained for case 2 in Table 2 via compact modeling and detailed modeling.

6.2. Correlations for the Coefficients as a Function of the Geometric and Operational Parameters

The previous section explained the different trends among the cases. Because the trends establish different behaviors for the parameters, obtaining correlations for the pressure loss coefficients as a function of the geometric and operational parameters is not easy. In order to obtain the correlations, certain restrictions were imposed on the coefficient values, in spite of knowing that the quality of the adjustment was going to be slightly reduced. Thermal designers will be interested in obtaining the values for the pressure loss coefficients for their geometric parameters on their own, which is why the correlations are more imperative than the values obtained by the iterative process.

The first restriction is to impose the value of zero on $\varphi_{z\,ext}$ for the cases that had a value of $\varphi_{z\,ext}$ greater than zero after the iterative process. As explained above, only two of the 81 cases turn out to have a positive value for $\varphi_{z\,ext}$. These two cases with a positive coefficient reduce the adjustment of the coefficient values predicted by the correlation to those obtained with the iterative process.

The second restriction is related to the value of $\varphi_{z\ int}$. It was verified that for the cases where $\varphi_{z\ ext}$ is smaller than zero (nearly all cases), the influence of the value for $\varphi_{z\ int}$ is strongly reduced, which makes it difficult for the iterative process to achieve good results for both coefficients. However, the shape of the flow pattern depends on the relation between $\varphi_{z\ ext}$ and $\varphi_{z\ int}$. For this reason, if one of the coefficients is fixed, it is possible to obtain a correct adjustment for the flow pattern through an adequate value for the other coefficient. Therefore, the value of $\varphi_{z\ int}$ is fixed to 100,000 (which is the value that this coefficient had for the two cases where $\varphi_{z\ ext}$ was originally greater than zero). This way, once the value of $\varphi_{z\ int}$ is fixed, it is easier for the iterative process to achieve a correct value for $\varphi_{z\ ext}$, and at the same time, the error for those two cases is minimized, as they have their original values for $\varphi_{z\ int}$. Eventually, the complexity of the model is also reduced from a compact model with three unknown pressure loss coefficients to a model with only two unknown coefficients.

With the above restrictions imposed on the values, the iterative process was applied again, this time only looking for $\varphi_{z\ ext}$ and φ_p, such that expression 10 was reduced to the following:

$$\begin{pmatrix} \phi_1 \\ \phi_2 \\ \phi_3 \\ \phi_4 \\ \phi_5 \end{pmatrix} - (A\ \ B)\begin{pmatrix} \varphi_{p\ 1} & \varphi_{p\ 2} & \varphi_{p\ 3} & \varphi_{p\ 4} & \varphi_{p\ 5} \\ \varphi_{z\ ext\ 1} & \varphi_{z\ ext\ 2} & \varphi_{z\ ext\ 3} & \varphi_{z\ ext\ 4} & \varphi_{z\ ext\ 5} \end{pmatrix} - D = 0 \quad (16)$$

With the new pressure loss coefficients, two correlations (one for φ_p and the other for $\varphi_{z\ ext}$) were obtained using DoE techniques [18]. Before they are presented, it is necessary to explain how they were used. The values of the parameters for the models had to be within the range shown in Table 1, and at the same time, their values had to be entered in a coded way. This means that instead of using the real value, it had to be transformed to the $(-1, 1)$ range according to expression 17 [18], where the real center value and the real low value are those that appear in Table 1 for the middle value (0) and for the minimum value (-):

$$Coded\ Value = \frac{Real\ Value - Real\ centre\ value}{Real\ centre\ value - Real\ low\ value} \quad (17)$$

The generic expression for both correlations can be written in a general form as [18]:

$$Y = b_0 + \sum_{i=1}^{k} b_i x_i + \sum_{i=1}^{k} b_{ii} x_i^2 + \sum\sum_{i<j} b_{ij} x_{ij} \quad (18)$$

where Y is the value obtained for the pressure loss coefficient through each correlation; x_i is the coded values for the geometric and operational parameters obtained through expression 17; and b_0, b_i, b_{ii} and b_{ij} for each correlation are the regression coefficients obtained through the DoE technique for the independent term, for each of the main effects, for the quadratic effects and for the significant interactions, respectively. k for each model is the number of main effects, quadratic terms or main interactions that must be included in order to attain a good fit for the regression models.

The correlations for φ_p and $\varphi_{z\ ext}$ are presented in Tables 4 and 5, respectively. The values of the regression adjusted coefficients (R^2) are 79% and 95%, respectively.

Table 4. Terms of the correlation for φ_p.

Factor	β_0	D	D_h/D	S	V	a	t
Value	445.48	−159.30	−52.03	22.94	−122.42	−22.02	257.08
Factor	ε	ξ	D-V	D-t	D-ε	D-ξ	S-D_h/D
Value	−307.91	230.03	84.87	155.50	61.81	53.03	−56.22
Factor	V-t	a-t	ε-t	ξ-t	ε-ξ		
Value	107.62	−60.56	−175.56	127.28	−127.16		

Table 5. Terms of the correlation for $\varphi_{z,\ ext}$.

Factor	β_0	D	D_h/D	S	V	a	t
Value	−771.53	519.09	142.53	−26.18	26.55	284.48	557.85
Factor	ε	ξ	D-D_h/D	D-t	D-ξ	a-D_h/D	t-D_h/D
Value	−74.12	175.85	75.16	−425.94	86.75	−116.31	−125.81
Factor	S-ε	V-ε	V-ξ	a-t	a-ε	t-ξ	t-t
Value	88.50	55.91	−53.75	−108.91	−57.81	−108.34	−193.59

6.3. Reliability of Compact Modeling

Having presented the correlations, the next step is to observe the results that are obtained through the use of the coefficients that are predicted by the correlations. With this aim, the flow magnitudes ($\overline{v_t}$, S, V_{min} and V_{max}) at the outlet cross-section of the domain are compared. On the one hand, the flow magnitudes obtained by the simulations with the pressure loss coefficients from the second iterative process are compared with those obtained by the simulations using the coefficients predicted by the correlations. On the other hand, the magnitudes obtained by the coefficient predictions of the correlations are compared with the magnitudes obtained by the detailed models in Bengoechea et al. [17] that are used as a reference. It must also be noted that the detailed model was experimentally validated in Bengoechea et al. [16].

The results for the first comparison are shown in Figure 7. From the figure, it can be concluded that using the values for φ_p and $\varphi_{z\ ext}$ obtained by the correlations from the compact model do not produce a big difference in the flow magnitudes with respect to using the coefficients obtained just after the iterative process. There are few samples that escape the ±20% of difference.

Once it was shown that the employed correlations had a reasonable influence on the flow magnitude predictions, these were compared with the flow magnitudes obtained by the detailed models in Bengoechea et al. [17]. This comparison is shown in Figure 8. From this figure, it can be seen that the flow magnitudes obtained by the compact model are satisfactory for $\overline{v_t}$, S and V_{max}. For these magnitudes, nearly all samples are inside the ±20% difference. However, for V_{min}, it seems that a considerable error is produced. Nevertheless, this error is not very significant if the velocity contours are analyzed. Figure 6 shows that in spite of there being a great difference between the minimum axial velocity values obtained by the compact model and those obtained by the detailed model, the approach to the velocity fields is excellent. Therefore, the results from the minimum axial velocities values are not significant. Moreover, in order to estimate whether the minimum velocity should be smaller, the reduced model presented in Bengoechea et al. [17] can be used.

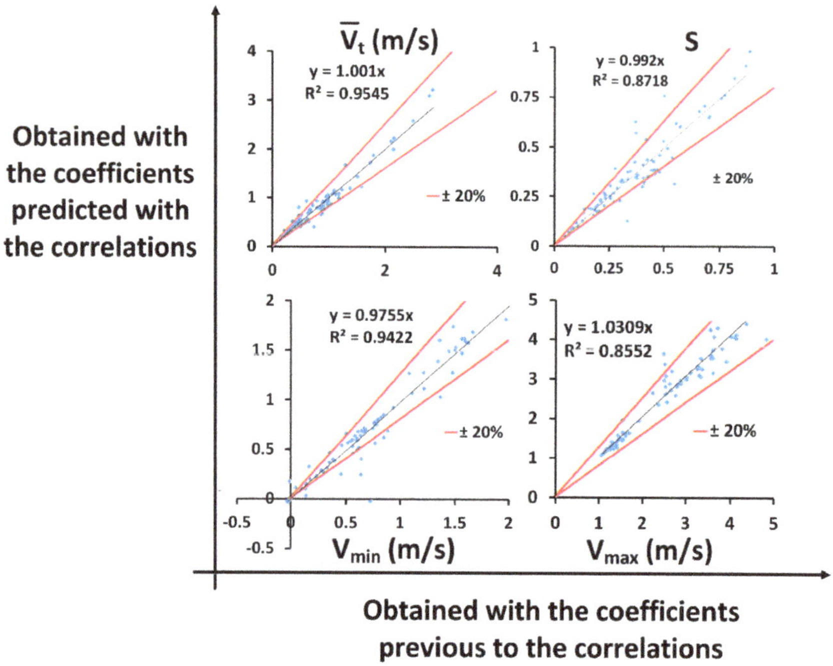

Figure 7. Flow magnitudes for the outlet cross section of the domain obtained with the coefficients predicted by the correlations vs. the flow magnitudes obtained with the coefficients previous to the correlations.

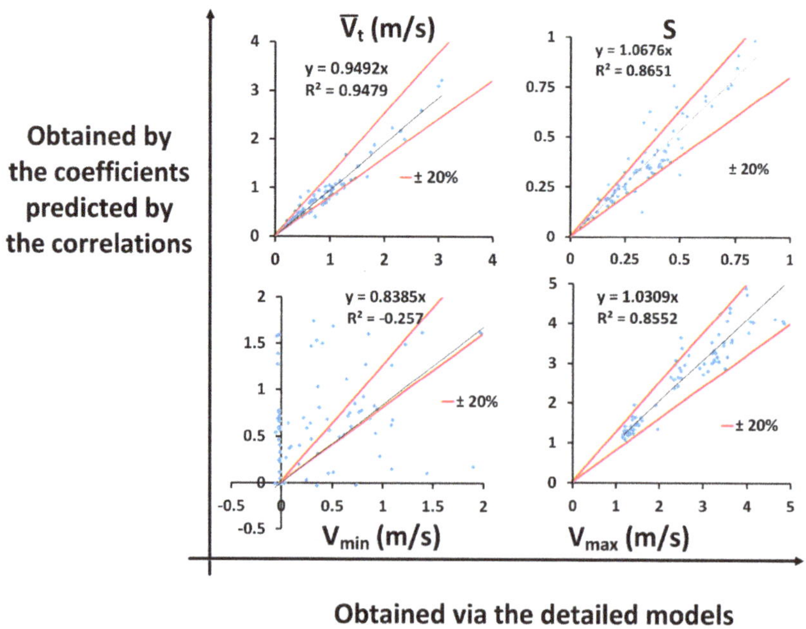

Figure 8. Flow magnitudes for the outlet cross-section of the domain obtained with the coefficients predicted by the correlations vs. the flow magnitudes obtained via the detailed models.

There is also interest in comparing how using the compact model improves the results that thermal designers attain without this model. Nowadays, thermal design is a challenge that is tackled by using pressure loss coefficients obtained from different handbooks, such as that by Idelchick [20]. The problem is that thermal designers use only one pressure loss coefficient. In order to show the improvements provided by the compact model, this pressure loss coefficient was calculated from Idelchick [20] for each of the 81 cases in Table 2. The geometries for each of the cases in Figure 2 were simulated by imposing the corresponding value of the calculated pressure loss coefficient on $\varphi_{z\ ext}$ and $\varphi_{z\ int}$, and $\varphi_p = 0$, which is the process used most frequently by thermal designers. In Figure 9, the flow magnitudes obtained via simulation for the outlet cross-section are compared with those obtained from the detailed models in Bengoechea et al. [17]. This figure shows that the predictions for each of the flow magnitudes ($\overline{v_t}$, S, V_{min} and V_{max}) made by using handbooks are much worse than the predictions obtained by the compact model presented in this work. In fact, the values for R^2 are clearly lower (unless for the maximum velocity). Furthermore, it is important to highlight that the correlation line is far from the ideal case of $y = x$, which means that in spite of having a good value for R^2 (as is the case for V_{max}), the error produced with respect to the detailed models is high for nearly all samples. As the figure indicates, for V_{max}, most of the samples are in the ±20% range of difference, in spite of having an R^2 of 0.93.

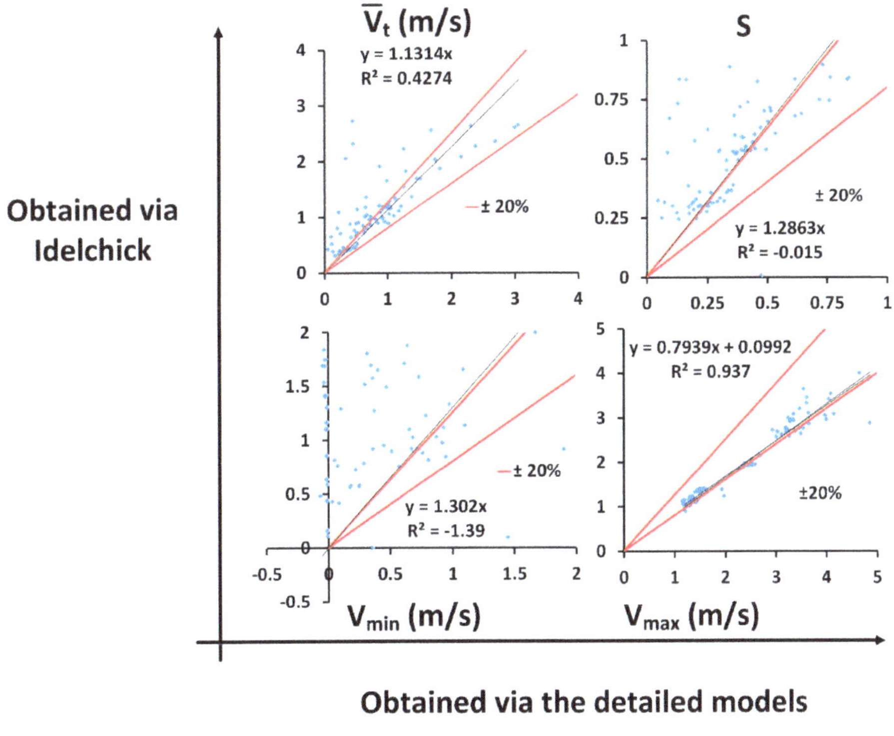

Figure 9. Flow magnitudes for the outlet cross-section of the domain obtained via Idelchick vs. the flow magnitudes obtained via the detailed models.

6.4. Application of the Compact Model with a Random Case

In order to check the correct working of the compact model and that the correlations are able to supply coefficients that approximate the flow pattern correctly, a simulation of a new detailed model case was performed. The modeling strategy is the same as that presented in Bengoechea et al. [17]. The geometric and operation parameters corresponding to this new case are presented in Table 6. Figure 10 presents the results obtained by the compact model (employing the pressure loss coefficients given by the correlations), by the detailed model, and by the coefficients provided by the Idelchick handbook [20]. As the figure shows, the prediction given by the compact model is closer to the detailed one than the prediction using the Idelchick handbook. The fit for the minimum and maximum axial velocities is better with the compact model that is introduced in this work, as is the adjustment for the tangential velocity. As the Idelchick handbook does not include pressure loss coefficients perpendicular to the flow direction, more tangential flow passes the screen when solving the flow by the coefficients given in this handbook. Therefore, the compact model presented in this work that consists of three directional pressure loss coefficients logically approximates a three-dimensional flow better.

It is necessary to note that the new case is supposed to be between those cases for which the Idelchick handbook is expected to provide a solution, as the level of the porosity of the screen is fixed to its higher value (see Table 6). This way, the value for the coefficient perpendicular to the flow direction tends towards a lower value. Therefore, through this new case, apart from showing that the compact model functions well, it was demonstrated that even for cases with high porosity, these correlations work better than the modeling currently employed.

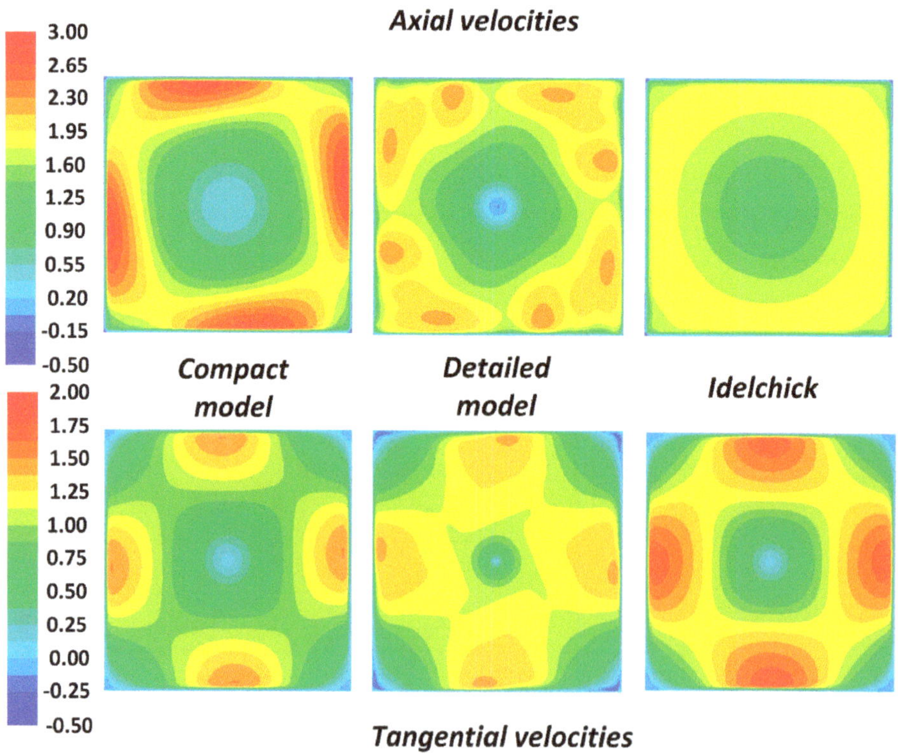

Figure 10. Axial and tangential velocity fields (m/s) for the outlet cross-section of the domain for the new case obtained by the compact model, by the detailed model and by Idelchick.

Table 6. Real and coded values for each parameter in the new case.

Parameter	D(mm)	D_h/D	S	a(mm)	ε (%)	ξ(1/mm)	t (mm)	V(m/s)
Real value	67	0.42	0.7	16	0.75	0.26	1.4	1.68
Coded value	−0.10	0.13	0.00	0.10	1	−0.66	−0.11	−0.09

7. Conclusions

A methodology for obtaining the pressure loss coefficients per unit length that define a compact model of an EMC screen based on a porous media approach was defined for a case in which the approaching flow is a fan-induced one. In addition, correlations predicting the values of the directional pressure loss coefficients as a function of the geometric and operational parameters were introduced. It is concluded that using this compact model offers thermal designers the following advantages with respect to using a detailed model or the modeling proposed by the various handbooks on pressure loss coefficients.

One advantage is that there is a very significant reduction in the time needed to create the geometry and the mesh, as it is not necessary to generate the geometries of the pores. Between 4 and 24 h is needed to create a detailed model, while no more than 10 min (less than 4% of the initial time) is required to create the compact model.

A second advantage is that as a consequence of the decrease in the number of elements that the mesh has, the time needed to converge the simulations is drastically reduced. For a detailed model, depending on the geometry, anywhere between 5 h and several weeks is needed to achieve a good level of convergence, whereas with a compact model, thermal designers have a reliable approximation of the flow pattern in 5 min (less than 1.7% of the initial time).

Therefore, the compact model presented in this work provides a quick and reliable approximation of the flow pattern that is superior to the approximations obtained by the modeling proposed in various handbooks. This approach permits electronic components to be placed into the 3D domain in such a way that a good prediction of the flow close to each component will be obtained quickly.

Author Contributions: Conceptualization, A.B. and R.A.; methodology, R.A.; software, A.B. and G.S.L.; validation, R.A., A.R. and J.C.R.; formal analysis, R.A., A.R. and J.C.R.; investigation, A.B. and R.A.; resources, R.A.; data curation, A.B.; writing—original draft preparation, A.B.; writing—review and editing, A.B., G.S.L., R.A., A.R. and J.C.R.; visualization, A.B., G.S.L., R.A., A.R. and J.C.R.; supervision, R.A.; project administration, R.A.; funding acquisition, R.A. All authors have read and agreed to the published version of the manuscript.

Funding: This research was funded by Eusko Jaurlaritza-Gobierno Vasco, Spain through Proyecto PI2011-30.

Institutional Review Board Statement: Not applicable.

Informed Consent Statement: Not applicable.

Data Availability Statement: Not applicable.

Acknowledgments: This research was funded by Eusko Jaurlaritza-Gobierno Vasco, Spain through Proyecto PI2011-30. The support of Cátedra Fundación Antonio Aranzábal-Universidad de Navarra is also gratefully acknowledged.

Conflicts of Interest: The authors declare no conflict of interest.

Nomenclature

a	distance between the fan and the EMC screen (mm)
A	area (m^2)
D	diameter of the fan and size of the channel (mm)
D_h	hub diameter (mm)
k	turbulent kinetic energy (m^2/s^2)
L_c	characteristic length (m)
P	pressure (N/m^2)
S	Swirl number (-)
t	thickness of the screen (mm)
$U_{i,j}$	velocity components (m/s)
v	velocity(m/s)
$\overline{v_t}$	area weighted average of the tangential velocity (m/s)
V	mean axial velocity (m/s)
V_a	local axial velocity (m/s)
V_t	mean tangential velocity (m/s)
Greek symbols	
β	regression coefficient (-)
ε	porosity of the screen (%)
ε	rate of dissipation of the turbulent kinetic energy (m^2/s^3)
φ	directional pressure loss coefficient (-)
μ	dynamic viscosity (N·s/m^2)
μ_t	turbulent viscosity (N·s/m^2)
ρ	density (kg/m^3)
ξ	ratio of perimeter to area of the pores of the screen (1/mm)

References

1. Sharma, C.S.; Tiwari, M.K.; Zimmermann, S.; Brunschwiler, T.; Schlottig, G.; Michel, B.; Poulikakos, D. Energy efficient hotspot-targeted embedded liquid cooling of electronics. *Appl. Energy* **2015**, *138*, 414–422. [CrossRef]
2. Birbarah, P.; Gebrael, T.; Foulkes, T.; Stillwell, A.; Moore, A.; Pilawa-Podgurski, R.; Miljkovic, N. Water immersion cooling of high power density electronics. *Int. J. Heat Mass Transf.* **2020**, *147*, 118918. [CrossRef]
3. Liu, Y.; Han, X.; Shen, C.; Yao, F.; Zhang, M. Experimental Study on the Evaporation and Condensation Heat Transfer Characteristics of a Vapor Chamber. *Energies* **2018**, *12*, 11. [CrossRef]
4. Estes, K.A.; Mudawar, I. Comparison of Two-Phase Electronic Cooling Using Free Jets and Sprays. *J. Electron. Packag.* **1995**, *117*, 323–332. [CrossRef]
5. Carneiro, M.V.P.; De Oliveira, P.A.; Barbosa, J.R. A compact refrigeration system based on multijet sprays for electronics thermal management. *Exp. Therm. Fluid Sci.* **2018**, *97*, 180–191. [CrossRef]
6. Hill, B.; Hill, C. Effects of Electronic Enclosure Layout on Fan Performance. In Proceedings of the ASME Winter Annual Meeting, Dallas, TX, USA, 25–30 November 1990. 90-WA/EEP-6.
7. Grimes, R.; Davies, M.; Punch, J.; Dalton, T.; Cole, R. Modeling Electronic Cooling Axial Fan Flows. *J. Electron. Packag.* **2000**, *123*, 112–119. [CrossRef]
8. Lin, S.C.; Chou, C.A. Blockage effect of axial-flow fans applied on heat sink assembly. *Appl. Therm. Eng.* **2004**, *24*, 2375–2389. [CrossRef]
9. Swim, W.B. An Experimental Study of the Effects of Inlet Plenum Walls on Axial Fan Performance. *ASHRAE Trans.* **2005**, *111*, 993–999.
10. Baniasadi, E.; Aydin, M.; Dincer, I.; Naterer, G. Computational Aerodynamic Study of Automotive Cooling Fan in Blocked Conditions. *Eng. Appl. Comput. Fluid Mech.* **2013**, *7*, 66–73. [CrossRef]
11. Antón, R.; Bengoechea, A.; Rivas, A.; Ramos, J.C.; Larraona, G.S. Performance of Axial Fans in Close Proximity to the Electromagnetic Compatibility Screens. *J. Electron. Packag.* **2012**, *134*, 011004. [CrossRef]
12. Tanner, P.; Gorman, J.; Sparrow, E. Flow–pressure drop characteristics of perforated plates. *Int. J. Numer. Methods Heat Fluid Flow* **2019**, *29*, 4310–4333. [CrossRef]
13. Miguel, A.F. Characterization of Fluid Flow through Perforated Plates. *J. Porous Media* **2019**, *22*, 1439–1448. [CrossRef]
14. Bejan, A.; Kim, S.J.; Morega Al, M.; Lee, S.W. Cooling of stacks of plates shielded by porous screens. *Int. J. Heat Fluid Flow* **1995**, *16*, 16–24. [CrossRef]
15. Brundrett, E. Prediction of Pressure Drop for Incompressible Flow Through Screens. *J. Fluids Eng.* **1993**, *115*, 239–242. [CrossRef]
16. Bengoechea, A.; Antón, R.; Larraona, G.S.; Rivas, A.; Ramos, J.C.; Masip, Y. PIV measurements and a CFD benchmark study of a screen under fan-induced swirl conditions. *Int. J. Heat Fluid Flow* **2014**, *46*, 43–60. [CrossRef]

17. Bengoechea; Anton, R.; Larraona, G.; Ramos, J.C.; Rivas, A. Influence of Geometrical Parameters in The Downstream Flow of A Screen Under Fan-Induced Swirl Conditions. *Eng. Appl. Comput. Fluid Mech.* **2014**, *8*, 623–638. [CrossRef]
18. Preece, D.A.; Montgomery, D.C. Design and Analysis of Experiments. *Int. Stat. Rev.* **1978**, *46*, 120. [CrossRef]
19. Baelmans, M.; Meyers, J.M.; Nevelsteen, K. Flow modeling in air-cooled electronic enclosures. In Proceedings of the Nineteenth Annual IEEE Semiconductor Thermal Measurement and Management Symposium, San Jose, CA, USA, 11–13 March 2003. [CrossRef]
20. Idelchik, I.E. *Handbook of Hydraulic Resistance*; Jaico Publishing House: Munbai, India, 2003.
21. Kordyban, T. Fan Swirl and Planar Resistances Don't mix. In Proceedings of the 9th International Flotherm User Conference, Orlando, FL, USA, 18–19 October 2000.
22. Anton, R.; Jonsson, H.; Moshfegh, B. Detailed CFD modeling of EMC screen for radio base stations: A benchmark study. *IEEE Trans. Compon. Packag. Technol.* **2007**, *30*, 754–763. [CrossRef]
23. Antón, R.; Jonsson, H.; Moshfegh, B. Compact CFD modelling of EMC screen for radio base stations: A porous media approach and a correlation for the directional loss coefficients. *IEEE Trans. Compon. Packag. Technol.* **2007**, *30*, 875–885. [CrossRef]
24. Antón, R.; Jonsson, H.; Moshfegh, B. Detailed CFD modelling of EMC screen for radio base stations: A conjugate heat trans-fer problem. *Int. J. Heat Exch.* **2007**, *8*, 95–116.
25. Anton, R.; Jonsson, H.; Moshfegh, B. Detailed CFD Modelling of EMC Screens for Radio Base Stations: A Parametric Study. *IEEE Trans. Compon. Packag. Technol.* **2009**, *32*, 145–155. [CrossRef]
26. Nevelsteen, K.; De Troch, K.; Mesbah, M.; Nelemans, W.; Baelmans, M. Screen characterization under fan induced swirl conditions. *IEEE Trans. Components Packag. Technol.* **2006**, *29*, 385–394. [CrossRef]
27. Najam, F.; Yu, Y.S. Compact Model for L-Shaped Tunnel Field-Effect Transistor Including the 2D Region. *Appl. Sci.* **2019**, *9*, 3716. [CrossRef]
28. Li, S.; Nezami, M.S.; Rolston, D.; Liboiron-Ladouceur, O. A Compact High-Efficient Equivalent Circuit Model of Mul-ti-Quantum-Well Vertical-Cavity Surface-Emitting Lasers for High-Speed Interconnects. *Appl. Sci.* **2020**, *10*, 3865. [CrossRef]
29. Colangelo, A.; Guelpa, E.; Lanzini, A.; Mancò, G.; Verda, V. Compact Model of Latent Heat Thermal Storage for Its Integra-tion in Multi-Energy Systems. *Appl. Sci.* **2020**, *10*, 8970. [CrossRef]
30. Beér, J.; Chigier, N. *Combustion Aerodynamics*; Robert, E., Ed.; Krieger Publishing Company Inc.: Malabar, FL, USA, 1983.
31. Launder, B.E.; Spalding, D.B. *Lectures in Mathematical Models of Turbulence*; Academic Press: London, UK, 1972.
32. Launder, B.; Spalding, D. The numerical computation of turbulent flows. *Comput. Methods Appl. Mech. Eng.* **1974**, *3*, 269–289. [CrossRef]
33. ANSYS FLUENT®13. In *Documentation*; Ansys, Inc.: Canonsburg, PA, USA, 2013.
34. Best Practice Guidelines. *European Research Community on Flow, Turbulence and Combustion, Version 1.0*; Best Practice Guidelines: Bushey, UK, 2000.

applied sciences

MDPI

Article

Particle Image Velocimetry Measurements of the Flow-Diverting Effects of a New Generation of the eCLIPs Implant for the Treatment of Intracranial Bifurcation Aneurysms

Sina G. Yazdi [1,*], Daniel Mercier [1], Renee Bernard [1], Adam Tynan [1] and Donald R. Ricci [1,2]

[1] Evasc Medical Systems, Vancouver, BC V6H 1C3, Canada; mercier@evasc.com (D.M.); bernard@evasc.com (R.B.); tynan@evasc.com (A.T.); ricci@mail.ubc.ca (D.R.R.)
[2] Department of Medicine, University of British Columbia, Vancouver, BC V5Z 1M9, Canada
* Correspondence: yazdi@evasc.com

Received: 7 November 2020; Accepted: 30 November 2020; Published: 3 December 2020

check for updates

Abstract: Flow diverters (FDs) for the endovascular treatment of intracranial aneurysms are effective for sidewall aneurysms, but their use at a bifurcation is problematic because FDs only partially cover the aneurysm neck and impede flow into a daughter branch; they are thus not employed routinely in this anatomy. eCLIPs was developed as a non-tubular implant to completely cover the neck of an aneurysm and serve as a coil retention device necessary for the adequate treatment of wide-neck bifurcation aneurysms. eCLIPs has shown some flow diversion effects in bifurcation anatomy but not equal to those exhibited by clinically accepted flow diverters in sidewall anatomy. A new generation of eCLIPs implant, the eCLIPs bifurcation flow diverter (eBFD), with higher metal coverage, was developed to achieve a similar flow diversion as a Pipeline Embolization Device (PED), a prototypical FD. Particle image velocimetry was used to capture the fluid dynamics and velocity reduction within silicone aneurysm replicas. A circulatory mimicking loop was developed to circulate the flow through the silicone models. All generations of eCLIPs implants had some flow-diverting effect, with increasing metal coverage density of the implant proportionately increasing the flow diversion effect. The eBFD, with a metal density of 35%, showed greater flow diversion than PED, with 30% metal density, for bifurcation anatomy. The eBFD showed similar reduction of flow in a bifurcation anatomy to PED in a sidewall, both sufficient to permit early thrombosis of the aneurysm. Thus, the eBFD can potentially provide sufficient flow diversion for the treatment of bifurcation aneurysms to avoid adjunctive coiling.

Keywords: particle image velocimetry; flow diverter; aneurysm; bifurcation; hemodynamic; velocity reduction; coil

1. Introduction

A brain aneurysm is a localized enlargement of an intracranial artery, usually due to damaged smooth muscle cells and elastin in the arterial layers [1] but often exacerbated by localized perturbation of blood flow. The Brain Aneurysm Foundation reported that the worldwide annual mortality due to intracranial aneurysms is 500,000, with approximately 50% of these patients being younger than 50 years old [2]. Common treatments for intracranial aneurysms include embolic coiling and stent-assisted coiling, but these are associated with frequent recurrence, especially for wide-neck aneurysms [3,4]. Stent-assisted coiling prevents coil protrusion into the parent artery lumen, a particular problem with wide-neck aneurysms, and reduces the recurrence risk [5]. Intra-aneurysmal coils act by initiating and propagating thrombus that ultimately cause the aneurysm to heal by thrombotic occlusion.

An emerging method to treat brain aneurysms is the use of flow diverter (FD) devices. Typically, FDs are tubular stents made of high-density braided mesh that minimizes blood perfusion through the stent wall into the aneurysm, producing blood stasis within the aneurysm that facilitates thrombus and thereby mitigates the need for coils [6]. All available FD devices in the market are designed for the treatment of sidewall aneurysms, but many aneurysms occur at a bifurcation. Y stenting is occasionally performed at a bifurcation to secure the coil and provide better neck coverage by deploying one stent from the parent artery to one daughter branch and then telescoping a second stent to the other (contralateral) branch [7,8]. FDs are not well suited to this anatomy because delivery methods preclude complete bridging of the neck. A FD is implanted from the parent artery to one daughter branch, providing a "neck narrowing" function to retain coils while also partially redirecting flow away from the aneurysm. Both Y stenting and FDs result in only partial coverage of the neck of the bifurcation, impede access to side branches, and, in the case of FDs, reduce the flow into contralateral side branches with the risk of hypoperfusion [9]. Another method for treating intracranial bifurcation aneurysms is with intrasaccular devices (e.g., WEB, Woven EndoBridge, Microvention, Aliso Viejo, CA, USA) [10]. While these devices disrupt intrasaccular flow to initiate thrombus, they do not reduce flow entering the neck of the aneurysm, a factor that can result in shape modification and compaction of intrasaccular contents [11,12].

eCLIPs (eCLIPs^{TM}, eVasc Neurovascular, Vancouver, BC, Canada) implants are semi-circumferential but non-tubular devices with two sections (Table 1): an anchor segment that is placed into one branch, and a leaf segment that crosses the neck of a bifurcation from one side branch to the other (contralateral) to effect coil retention without blocking the flow to the branches [13]. Peach et al. [14] demonstrated the FD characteristics of the eCLIPs leaf section using computational flow dynamics (CFD). They compared the eCLIPs flow diversion against that of clinically accepted FDs, e.g., the Pipeline Embolization Device (PED, Medtronic, Minneapolis, MN, USA), in a bifurcation geometry. They concluded that eCLIPs is 10% less effective than braided mesh in reducing the inflow and the jet into the aneurysm, possibly due to the larger pore size, but does not impede flow into either side branch. Different numerical and experimental hemodynamic studies had shown that reducing the pore size leads to a greater flow diversion effect [15–17]. Particle image velocimetry (PIV) is an optical measurement technique that has been widely used for in vitro assessment of the intracranial aneurysm hemodynamics and flow diversion effect [16–18].

The current state of the art in the endovascular treatment of intracranial aneurysms does not offer a universally reliable and safe method to treat bifurcation aneurysms but suggests that FDs in some form may offer a solution. The eCLIPs implant has had three iterations: an early "proof of principle" stainless steel version [19] and two nitinol versions, the earlier (eCLIPs bifurcation remodelling system, eBRS) having FD characteristics as described by Peach et al. [14] and the later having improved deliverability (eCLIPs electrolytic bifurcation, eB) [20]. A fourth generation of eCLIPs implant (eCLIPs bifurcation flow diverter, eBFD) was developed with increased metal coverage (density) to enhance the flow diversion effect of the eCLIPs implants at a bifurcation. The objective of this study was to use PIV to assess the degrees of flow diversion of the successive generations of eCLIPs implant, particularly the fourth generation (eBFD), and to compare the results against those of a clinically accepted FD, PED (Table 1), which was tested in a sidewall and bifurcation aneurysm configuration. Enhanced flow diversion by the eBFD may be sufficient for healing a bifurcation aneurysm without the need for adjunctive coiling.

Table 1. Illustration and general characteristics of the three versions of eCLIPs and Pipeline studied using particle image velocimetry (PIV) analysis; the Pipeline device image was adapted from Dholakia et al. [21].

Series	Implant Figure		Metal Coverage (Density)	Common Geometry Implanted	Purpose of Use
eCLIPs Series	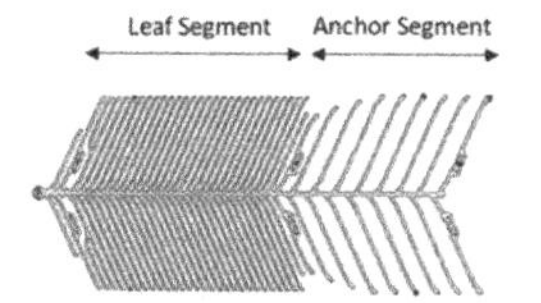	**eBRS**	26%	Bifurcation aneurysm	Coil retention
	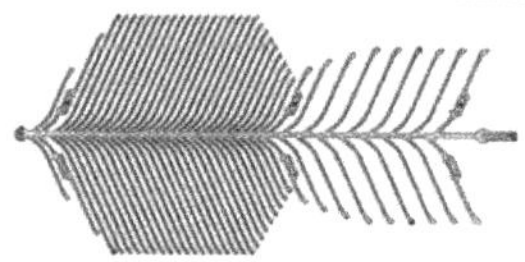	**eB**	21%	Bifurcation aneurysm	Coil retention
	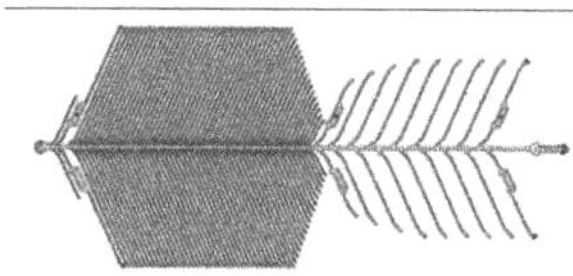	**eBFD**	35%	Bifurcation aneurysm	Flow diversion
Pipeline (PED)			30%	Sidewall aneurysm	Flow diversion

2. Materials and Methods

Two arterial phantoms were used in this study. First, an idealized 1:1 scale compliant phantom of the basilar artery with posterior cerebral branches was designed with an aneurysm at the bifurcation (Figure 1a). Another 1:1 scale model of the basilar artery with a sidewall aneurysm was designed (Figure 1b). The phantoms were manufactured from Silicone Sylgard-184 (Dow Corning Inc. Midland, MI, USA) at room temperature (23 °C) (Figure 1c). The lost core casting method developed by Yazdi et al. [22] was used for phantoms manufacture; the details of the moulding process can be found in their study. It was reported that the Sylgard-184 has a Young's modulus of 1.32 MPa if it is cured at room temperature [23,24]; thus, it was not measured in this study.

In total, 7 experiments were conducted and 4 implants were tested. Cases 2 and 3 (generations 2 and 3) were eCLIPs coil assisting implants commercially available on the market. Case 4 (generation 4) was a revision developed with higher metal coverage. A schematic view of an eCLIPs implant in bifurcation configuration is shown in Figure 2a. Since the PED implant is the prototypical FD, its measure of flow diversion was used as a reference to compare the eCLIPs' results. The PED was implanted in both bifurcation (Case 5) (Figure 2b) and sidewall (Case 7) configurations (Figure 2c). Cases 1 and 6 were the silicone models without any implant and served as baselines to assess the flow diversion effect.

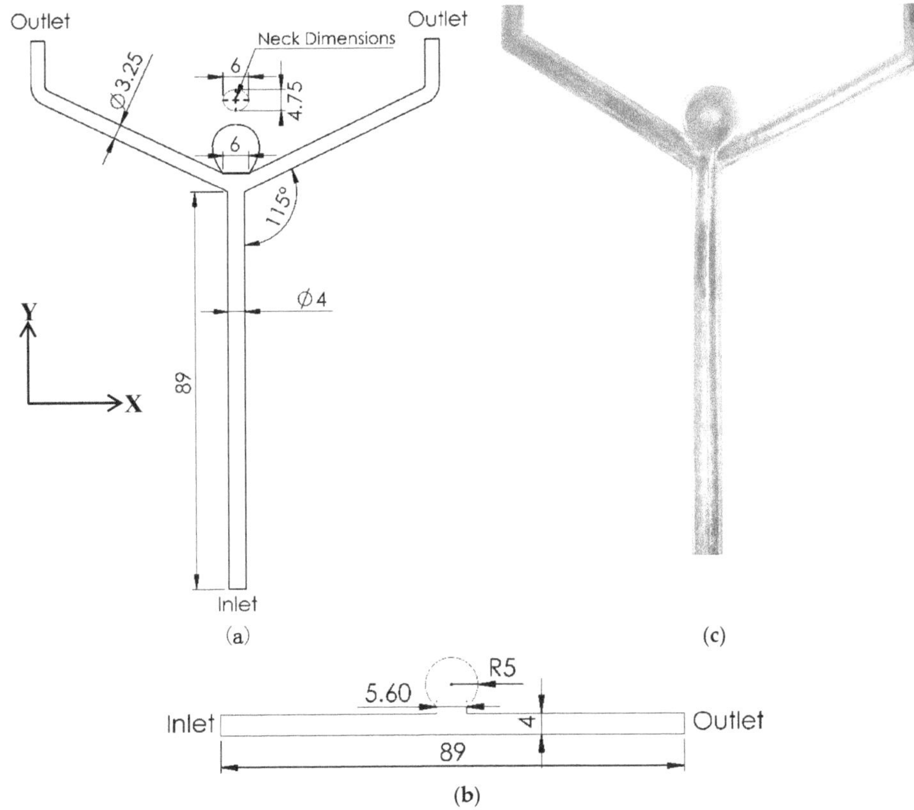

Figure 1. (**a**) Bifurcation aneurysm model drawing, (**b**) sidewall model drawing, and (**c**) sample image of the manufactured silicone model; length units are in mm.

A fluid circuit was developed to run the flow through the phantoms (Figure 3). A solution of 47.38% water, 36.94% glycerine, and 15.68% sodium iodide was used as a blood analogue [25]. The circuit consisted of a steady pump, magnetic-inductive flowmeter (SM 4000, IFM Inc., Daimlerring, Germany), phantom box, reservoir, and header tank. The header tank was used to mimic the impedance of the smaller arteries and capillaries. To eliminate the vibration in the fluid line and aneurysm due to the pump propulsion, the fluid was pumped to a reservoir (Figure 3) and gravity-fed to the model. A weir was designed in the reservoir to keep the fluid level constant. To eliminate optical distortion during the PIV experimentation, the working fluid should have the same refractive index as the silicone phantoms. The phantom box was also filled with the same working fluid to mimic the surrounding tissue. This solution matched the refractive index of the silicone Sylgar-184 (n = 1.414), which was verified by placing a checkerboard behind the phantom and capturing images to check for distortion of the grids. The solution had dynamic viscosity of 4.31 cP, which is in the range of human blood viscosity [25]. The volumetric flow rate in the basilar artery can reach up to 300 mL/min at peak systole [17,26]. Since it was expected to see the highest velocity and blood protrusion into the aneurysm at peak systole, all the experiments were conducted at a 300 mL/min steady flow. Since there are no published data on the shear rate threshold for successful flow diversion and thrombosis formation in a bifurcation aneurysm, the current study only mapped the velocity field in the aneurysm sac and discussed the flow diversion based on the velocity reduction.

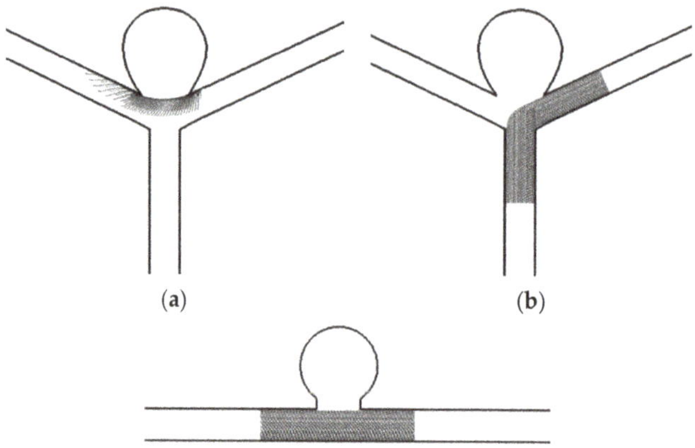

Figure 2. Schematic views of the deployed implants in different configurations: (**a**) eCLIPs in bifurcation, (**b**) PED in bifurcation, and (**c**) PED in sidewall configuration.

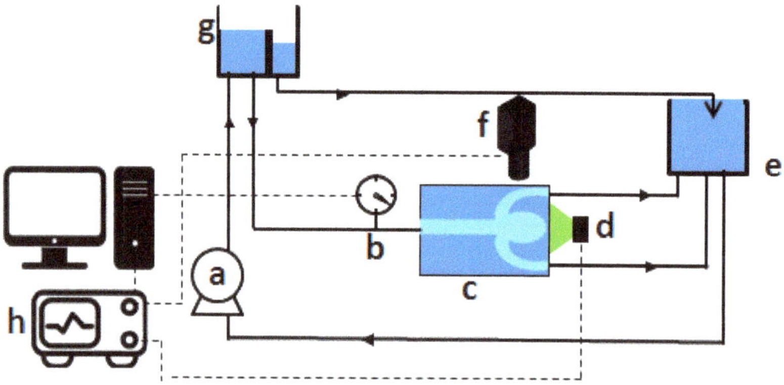

Figure 3. Schematic view of the fluid circuit and PIV setup: (**a**) steady pump, (**b**) magnetic-inductive flowmeter, (**c**) phantom box, (**d**) LED light source, (**e**) header tank, (**f**) PIV camera, (**g**) reservoir (header tank with weir), (**h**) control platform.

Planar PIV was used to capture the fluid dynamics within the aneurysm dome. A high-power LED (ILA-5150 GmbH, Aachen, Germany) light source was used to illuminate the region of interest. Fiber optics connected to the LED head transferred the illumination to the light sheet generator composed of cylindrical lenses. The adjusted light sheet had a thickness of 1 mm and was aligned with the central coronal plane of the aneurysm for which all the results were captured. An ILA μEYE PIV camera (ILA-5150 GmbH, Aachen, Germany) with 2048 × 2048 pixels was used to capture the particle images. The camera was equipped with a 35 mm lens installed on a 40 mm C-mount extension. This allowed the lens to focus on objects at shorter distances than would be possible without a spacer, thus increasing the magnification of the image on the camera sensor. The flow was seeded polyamide particles with a nominal diameter of 5 μm. An ILA network synchronizer was used to fire the LED in double-pulsed mode and synchronize the illumination and capture events.

Light reflection from the silicone model wall and stationary objects can cause an error in displacement vector calculation. Thus, to increase the signal-to-noise ratio before cross-correlation, the average intensity background was subtracted from the images. Gaussian window weighting was applied to images prior to the correlation. Preprocessing of the images eliminated the peak-locking effect, which is a cause of bias in the PIV analysis and increases the velocity uncertainty. Cross-correlation was used to compute the displacement vector field. For all analyses, 100 image pairs were captured and used for the analysis of the ensemble-averaged vector field. Almost no interpolated vector was observed in the ensemble averaged vector field. A combination of the mentioned conditions ensured low uncertainty of the calculated velocity field. Iterative window refinement was used for grid generation with start and final window dimensions of 225 × 225 and 48 × 48 pixels, respectively, with 50% grid overlapping. Normalized median testing of the 8 neighboring grids was used to validate the computed vector field. PIVview2C software was used for the analysis of the images. The mapped velocity data were plotted using Tecplot 360 software.

3. Results

Figure 4 shows the velocity contour plots of all cases with and without implants in bifurcation and sidewall configurations. Due to the difference in peak velocity between the control case and the cases with implants, a logarithmic scale color map was used to enhance the comparison. In general, the figures show the effect of higher metal coverage on flow diversion by progressively reducing velocity within the aneurysm.

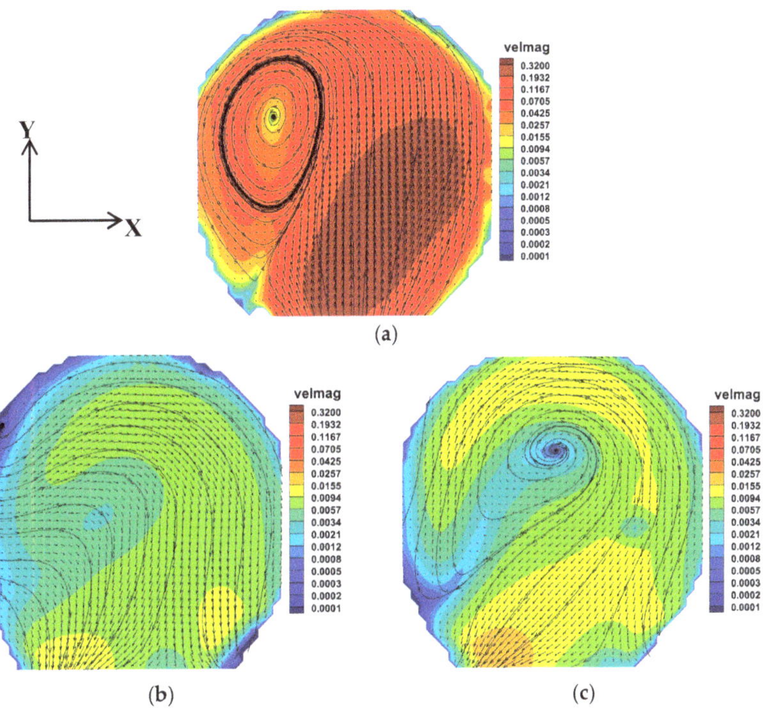

Figure 4. *Cont.*

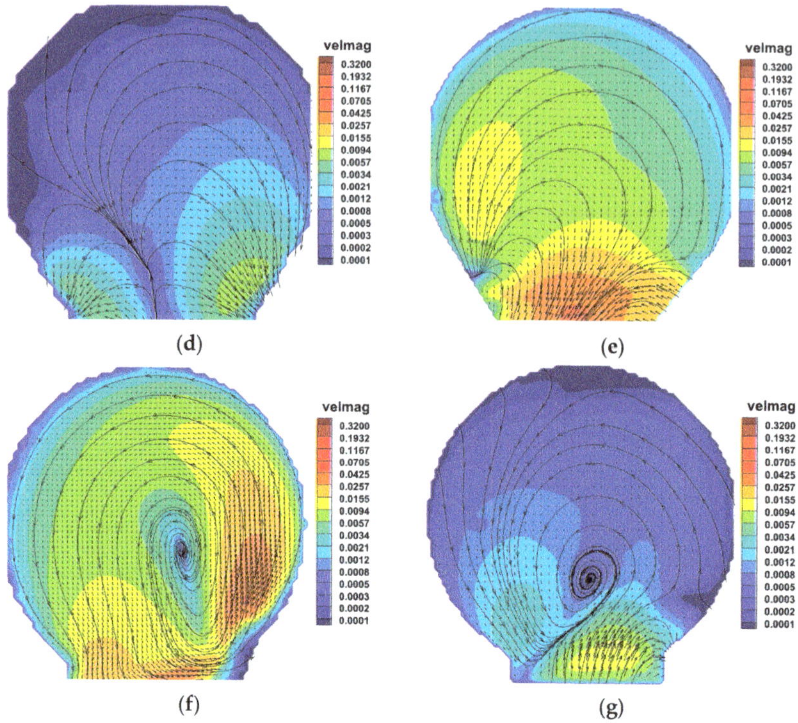

Figure 4. Velocity colour mapping and flow streamlines for different implants in sidewall and bifurcation geometry; note that logarithmic color mapping was used to enhance the comparison of contours. (**a**) Case 1, bifurcation, no implant, (**b**) Case 2, bifurcation, eBRS, (**c**) Case 3, bifurcation, eB, (**d**) Case 4, bifurcation, eBFD, (**e**) Case 5, bifurcation, PED, (**f**) Case 6, sidewall, no implant, (**g**) Case 7, sidewall, PED.

Table 2 shows the mean and maximum velocity measurements for each case and the percentage of reduction compared with the control case. As the metal coverage density increases, the mean and maximum velocities are reduced. The highest velocity reduction was observed in Case 4, the eCLIPs fourth-generation eBFD, which had the highest density.

Table 2. Peak and average aneurysm velocities and the percentage of reduction for each case at 300 mL/min volumetric flow rate.

Case	Density%	Max. Velocity (m/s)	Max Velocity Reduction (%)	Mean Velocity (m/s)	Mean Velocity Reduction (%)
Bifurcation					
Case 1, no implant	N/A	0.320	N/A	0.127	N/A
Case 2, eBRS (Generation 2)	26	0.0172	95	0.0049	96
Case 3, eB (Generation 3)	21	0.0260	92	0.0077	94
Case 4, eBFD (Generation 4)	35	0.0100	97	0.0014	98
Case 5, PED 3.75 mm × 18 mm	30	0.0380	88	0.0075	94
Sidewall					
Case 6, no implant	N/A	0.0500	N/A	0.0096	N/A
Case 7, PED 3.75 mm × 18 mm	30	0.0144	84	0.0013	86

Figure 4a shows the bifurcation aneurysm without an implant (control case). A jet was observed that entered the aneurysm predominantly from the distal end of the neck. The maximum velocity obtained was 0.32 m/s, which is significantly close to the theoretical velocity of 0.39 m/s calculated in the parent artery given the volumetric flow rate and model dimensions stated earlier. A considerable recirculation can be seen in the velocity map, with the core located near the top of the dome. For the eBRS implant (Figure 4b), the velocity was reduced across the dome, and there was no evidence of recirculation. The eB implant (Figure 4c) showed a similar flow pattern to the eBRS. However, a recirculation was observed, relatively weaker than that in the control case. The recirculation was counter-clockwise in the control case, while a clockwise recirculation was observed in the eB case. Amongst all the implants in the bifurcation configuration, the eBFD showed the best flow diversion performance and velocity reduction (Figure 4d). The maximum and mean velocity were reduced by 97% and 98%, respectively (Table 2). The velocity was significantly reduced at the dome, with no evidence of recirculation, and the highest velocity was observed at the sides of the neck.

The PED in the bifurcation geometry (Figure 4e) showed a lesser level of flow diversion than the eBFD, and the mean velocity reduction was the same as that of eB (Table 2). The flow entered the aneurysm from the distal end of the neck with a relatively high core velocity.

In the sidewall geometry without an implant, flow entered the aneurysm from the distal end of the neck, and a jet with a core velocity of 0.05 m/s was observed (Figure 4f). The flow circulated at the center of the aneurysm. Of note is that the velocity entering the aneurysm at the bifurcation location (0.32 m/s, Table 2) was markedly higher than that entering at the sidewall location (0.05 m/s, Table 2). Deployment of the PED in the sidewall configuration (Figure 4g) led to a considerable reduction of the velocity in the aneurysm sac, particularly at the top of the dome. The velocity color mapping showed that the eBFD at a bifurcation (Case 4) resulted in a greater reduction in inflow velocity—mean and maximal velocity reductions of 97% and 98%, respectively—than did the PED at a sidewall (Case 7)—mean and maximal velocity reductions of 84% and 86%, respectively.

4. Discussion

Flow diverters have been accepted as standalone implants for the treatment of intracranial aneurysms in lieu of other conventional treatment methods, including clipping, embolic coil, and stent-assisted coiling, in particular for wide-neck aneurysms [27]. However, effective flow diversion by the implant for aneurysm healing is considerably dependent on the velocity reduction within the aneurysm sac [28]. The current study used a PIV optical fluid dynamics measurement technique to evaluate the flow diversion effect of a new generation of the eCLIPs flow diverter and compare the results to those of previous eCLIPs coil retaining implants and the PED flow diverter.

The deployment of all devices caused a flow-diverting effect in the aneurysm dome. However, the level of flow diversion and reduction of aneurysmal velocity compared with the baseline were different across the implants and configurations. The results showed that there is a strong correlation between metal coverage density and flow reduction. This finding was in agreement with previously published studies [15–17]. For instance, the eBRS and eB implants both reduced the velocity within the aneurysm, with metal densities of 26% and 21%, respectively.

Yu et al. [29] experimentally investigated the effect of stent porosity in reducing cerebral sidewall aneurysm flow using PIV. They reported that to effectively dampen the flow in an aneurysm, the stent porosity should be kept around 60% (40% metal coverage), and for a porosity of 64% (36% metal coverage), the velocity decreased by 98% in the aneurysm [29]. The eBFD was manufactured with 35% metal coverage. It had the highest flow reduction amongst the implants, with a mean flow reduction of 98%, concordant with results by Yu et al. [29]. Their observation that a 26% density yielded a mean velocity of 0.0048 m/s in the aneurysm is comparable to the mean velocity of 0.0049 m/s observed for the eBRS (also 26% density). Thus, Yu's data provide a measure of validity to the results obtained in this report.

A jet and large recirculation was observed in the sidewall anatomy (Figure 4f) with no device across the neck. A computational study by Castro et al. [30] showed similar fluid dynamics in a cerebral bifurcation without an implant at peak systole. As expected, the PED implant in sidewall configuration significantly reduced the flow in the aneurysm dome (Figure 4g)—a result of the high density of the braided mesh influencing the fluid dynamics in the aneurysm dome. The PED has been widely accepted as an effective flow diverter for the treatment of sidewall intracranial aneurysms. A review by Wong et al. [6] of PED flow diverters' clinical outcomes reported that the PED is an effective implant to treat wide-necked, fusiform, large, and giant unruptured aneurysms, with 5% of patients experiencing permanent major morbidity and mortality. However, they emphasized a lack of understanding of the flow diverter's effect at bifurcations.

While the PED has 30% metal coverage, as reported in the device's instruction for use, the PED in the bifurcation configuration showed higher velocity entering the aneurysm (mean velocity 0.0075 m/s) than either the eBRS (0.0049 m/s) or eBFD (0.0014 m/s) and effectively the same as the eB implant (0.0077 m/s) with 21% density (Table 2). The higher PED mean velocity in the bifurcation geometry over a device that fully bridges the neck, such as eCLIPs implants, may be explained by the PED bending at the transition between the parent artery and daughter branch, leading to mesh stretching and higher porosity. In addition, the PED in the bifurcation configuration only partially covers the aneurysm neck, potentially leading to a Venturi effect and high velocity at the neck. While not quantified, visual inspection showed an obvious, and expected, reduction of the outflow into the contralateral branch. This observation was highlighted by Tang et al. [9], who computationally modelled the blood flow in a bifurcation geometry pre- and post-implantation of a PED and showed a reduction of the flow rate into the contralateral branch, leading to hypoperfusion post stenting.

The velocity entering the aneurysm at the bifurcation location is markedly higher than that entering at a sidewall location, as might be expected due to the flow perturbation unique to a bifurcation geometry [11], comporting with the clinical observation that a water jet phenomenon at a bifurcation is the likely cause of the higher rate of recurrence at a bifurcation [31].

The eBFD flow diverter in the bifurcation shows a reduction of mean velocity of 98%, whereas the PED has a mean reduction of mean velocity of 86% in a sidewall configuration. Possibly more importantly, the mean velocity is 0.0014 m/s within a bifurcation aneurysm with an eBFD implant and is 0.0013 m/s for a PED implant at a sidewall. Mut et al. [28], using CFD simulation, studied the association of hemodynamic conditions and occlusion times after flow diversion in cerebral aneurysm. Their study reported that the threshold for early occlusion of the aneurysm (thrombosis at 3 months) is a reduction of aneurysm velocity to 0.013 m/s. Both the eBFD and PED exceeded this threshold in the bifurcation and sidewall locations, respectively, by a factor of 10. Recognizing that the PED is accepted in clinical use as a FD in sidewall applications with a mean velocity of 0.0013 m/s, the eBFD, with an aneurysm mean velocity of 0.0014 m/s, should provide the same security of early thrombus formation in a bifurcation aneurysm as PED confers in a sidewall aneurysm. The successful flow-diverting effect of the eBFD in a bifurcation aneurysm, however, would occur without obstructing the flow into either side branch.

While wall shear stress (WSS) is an important parameter in the analysis of aneurysm flow diversion effects, it was not calculated in this study due to aliasing and light reflection at the lumen wall that may lead to error in the cross-correlation algorithm. CFD analysis can be implemented in the future to calculate the WSS in three dimensions and high resolution. Kim et al. [32] numerically showed that the placement of the stents reduced the velocity and vortices in the aneurysm sac, which caused a reduction of the wall shear stress and shear rate. In the current study, we also observed velocity and recirculation area reductions in the aneurysm after implantation. Since the shear rate and WSS are directly proportional to the velocity gradient, it is expected that the shear rate and WSS stress are reduced, similar to the results by Kim et al. [32].

While blood exhibits non-Newtonian behavior for a shear rate lower than 100 1/s, the working fluid used in this study was Newtonian. Najjari et al. [33] reported that the refractive index of xanthan-gum-based non-Newtonian fluid cannot be matched easily by varying NaI as it considerably changes the fluid viscosity. Failing to perfectly match the refractive index causes optical distortion in PIV analysis and bias in the cross-correlation algorithm, leading to inaccurate results. Thus, a

Newtonian fluid was chosen as a blood analogue. Using a non-Newtonian blood analogue might yield improved outcomes. However, different experimental and numerical hemodynamic studies have used a Newtonian model and proved that it provides a good estimation of flow characteristics [34,35]. Valencia et al. [35] numerically investigated the effect of the non-Newtonian fluid assumption on blood flow dynamics in the internal carotid artery with a saccular aneurysm. The study reported that the non-Newtonian effect on wall shear stress was only important in high-velocity-gradient areas. However, both Newtonian and non-Newtonian models showed similar wall shear stress in the aneurysm area. Blood flow has pulsatile characteristics. The current study employed a steady flow for analysis, but this model should not mitigate the value of the results shown herein since different studies have shown that steady flow provides a strong indication of fundamental hemodynamic trends that are present under both steady and pulsatile conditions in the aneurysm [36,37].

All the results presented were captured for only the central coronal plane (XY plane) of the aneurysm. Planar PIV was used in this study to capture the fluid dynamics and could not capture the out-of-plane component of the velocity. While it was expected to observe a complex flow in the XZ plane, the out-of-plane behavior of the flow remains ambiguous for future studies. If the out-of-plane velocity and transverse secondary flow development are deemed necessary, stereo-PIV, tomographic PIV [38,39], or three-dimensional phase-contrast MRI can be utilized [40]. These may be employed in future studies to help verify the results presented herein.

5. Conclusions

eCLIPs has undergone three iterations for the treatment of intracranial bifurcation aneurysms, the first never achieving clinical uptake due to the delivery impediments associated with its balloon-expandable stainless steel design. The second generation, the eCLIPs eBRS, with self-expanding nitinol, showed a calculable flow diversion effect at a bifurcation. A third generation has improved delivery characteristics and compatibility with smaller catheters. A fourth generation, the eBFD, was developed to build upon these improvements and also to try to achieve a sufficient flow diversion effect to obviate the need for adjunctive coiling. This study used PIV to measure the flow-diverting hemodynamics, implied by reducing the velocity in the aneurysm sac, in successive eCLIPs versions in comparison with a clinically recognized flow diverter, the PED, used as a reference predicate. The results showed that for various eCLIPs iterations, increasing metal coverage density was proportional to a reduction of the mean and maximum velocity in the aneurysm sac. The PED was not only less effective as a flow diverter in the bifurcation configuration than the eBFD, but also less effective at a bifurcation than in a sidewall configuration. The eBFD (fourth generation) showed equivalent flow diversion at a bifurcation to PED in sidewall anatomy. Thus, it was concluded that the eBFD has the potential for sufficient flow diversion and thrombosis in a bifurcation aneurysm without the need for adjunctive coiling. This thesis must be validated by clinical studies. The study also demonstrated the efficacy of the PIV measurement technique in assessing implant flow diversion and associated hemodynamics.

Author Contributions: Conceptualization, S.G.Y., D.M., R.B., A.T., and D.R.R.; methodology: S.G.Y.; software, S.G.Y.; validation, S.G.Y., D.M.; investigation, S.G.Y.; writing—original draft preparation, S.G.Y.; writing—review and editing, S.G.Y., D.M., R.B., A.T. and D.R.R.; visualization, S.G.Y.; supervision, D.M., A.T., and D.R.R. All authors have read and agreed to the published version of the manuscript.

Funding: This research received no external funding.

Conflicts of Interest: The authors declare no conflict of interest.

References

1. Jung, K.-H. New Pathophysiological Considerations on Cerebral Aneurysms. *Neurointervention* **2018**, *13*, 73–83. [CrossRef] [PubMed]

2. Foundation, B.A. Statistics and Facts—Brain Aneurysm Foundation. Available online: https://bafound.org/about-brain-aneurysms/brain-aneurysm-basics/brain-aneurysm-statistics-and-facts/ (accessed on 9 October 2020).
3. Shapiro, M.; Becske, T.; Sahlein, D.; Babb, J.; Nelson, P.K. Stent-supported aneurysm coiling: A literature survey of treatment and follow-up. *Am. J. Neuroradiol.* **2012**, *33*, 159–163. [CrossRef] [PubMed]
4. Johnston, S.C.; Dowd, C.F.; Higashida, R.T.; Lawton, M.T.; Duckwiler, G.R.; Gress, D.R. Predictors of rehemorrhage after treatment of ruptured intracranial aneurysms: The Cerebral Aneurysm Rerupture After Treatment (CARAT) study. *Stroke* **2008**, *39*, 120–125. [CrossRef] [PubMed]
5. Wanke, I.; Forsting, M. Stents for intracranial wide-necked aneurysms: More than mechanical protection. *Neuroradiology* **2008**, *50*, 991–998. [CrossRef]
6. Wong, G.K.C.; Kwan, M.C.L.; Ng, R.Y.T.; Yu, S.C.H.; Poon, W.S. Flow diverters for treatment of intracranial aneurysms: Current status and ongoing clinical trials. *J. Clin. Neurosci.* **2011**, *18*, 737–740. [CrossRef]
7. Jankowitz, B.T.; Thomas, A.; Jovin, T.; Horowitz, M. Y stenting using kissing stents for the treatment of bifurcation aneurysms. *J. Neurointerv. Surg.* **2012**, *4*, 16–21. [CrossRef]
8. Darkhabani, Z.M.; Lazzaro, M.A.; Zaidat, O.O. Pericallosal artery aneurysm treatment using Y-configuration stent-assisted coil embolization: A report of four cases. *J. Neurointerv. Surg.* **2012**, *4*, 459–462. [CrossRef]
9. Tang, A.Y.S.; Chung, W.C.; Liu, E.T.Y.; Qu, J.Q.; Tsang, A.C.O.; Leung, G.K.K.; Leung, K.M.; Yu, A.C.H.; Chow, K.W. Computational Fluid Dynamics Study of Bifurcation Aneurysms Treated with Pipeline Embolization Device: Side Branch Diameter Study. *J. Med. Biol. Eng.* **2015**, *35*, 293–304. [CrossRef]
10. Arthur, A.S.; Molyneux, A.; Coon, A.L.; Saatci, I.; Szikora, I.; Baltacioglu, F.; Sultan, A.; Hoit, D.; Almandoz, J.E.D.; Elijovich, L. The safety and effectiveness of the Woven EndoBridge (WEB) system for the treatment of wide-necked bifurcation aneurysms: Final 12-month results of the pivotal WEB Intrasaccular Therapy (WEB-IT) Study. *J. Neurointerv. Surg.* **2019**, *11*, 924–930. [CrossRef]
11. De Vries, J.; Boogaarts, H.D.; Sørensen, L.; Holtmannspoetter, M.; Benndorf, G.; Turowski, B.; Bohner, G.; Derakhshani, S.; Navasa, C.; van Zwam, W.H. eCLIPs bifurcation remodeling system for treatment of wide neck bifurcation aneurysms with extremely low dome-to-neck and aspect ratios: A multicenter experience. *J. Neurointerv. Surg.* **2020**. [CrossRef]
12. Caroff, J.; Mihalea, C.; Ikka, L.; Ozanne, A.; Gallas, S.; Chalumeau, V.; Moret, J.; Spelle, L. O-009 WEB shape modification during follow-up: The bicêtre experience. *J. Neurointerv. Surg.* **2020**. [CrossRef]
13. Chiu, A.H.; De Vries, J.; O'Kelly, C.J.; Riina, H.; McDougall, I.; Tippett, J.; Wan, M.; De Oliveira Manoel, A.L.; Marotta, T.R. The second-generation eCLIPs Endovascular Clip System: Initial experience. *J. Neurosurg.* **2018**, *128*, 482–489. [CrossRef] [PubMed]
14. Peach, T.W.; Ricci, D.; Ventikos, Y. A Virtual Comparison of the eCLIPs Device and Conventional Flow-Diverters as Treatment for Cerebral Bifurcation Aneurysms. *Cardiovasc. Eng. Technol.* **2019**, *10*, 508–519. [CrossRef] [PubMed]
15. Augsburger, L.; Farhat, M.; Reymond, P.; Fonck, E.; Kulcsar, Z.; Stergiopulos, N.; Rüfenacht, D.A. Effect of flow diverter porosity on intraaneurysmal blood flow. *Clin. Neuroradiol.* **2009**, *19*, 204–214. [CrossRef] [PubMed]
16. Liou, T.M.; Li, Y.C. Effects of stent porosity on hemodynamics in a sidewall aneurysm model. *J. Biomech.* **2008**, *41*, 1174–1183. [CrossRef] [PubMed]
17. Roszelle, B.N.; Gonzalez, L.F.; Babiker, M.H.; Ryan, J.; Albuquerque, F.C.; Frakes, D.H. Flow diverter effect on cerebral aneurysm hemodynamics: An in vitro comparison of telescoping stents and the Pipeline. *Neuroradiology* **2013**, *55*, 751–758. [CrossRef]
18. Dennis, K.D.; Rossman, T.L.; Kallmes, D.F.; Dragomir-Daescu, D. Intra-aneurysmal flow rates are reduced by two flow diverters: An experiment using tomographic particle image velocimetry in an aneurysm model. *J. Neurointerv. Surg.* **2015**, *7*, 937–942. [CrossRef]

19. Marotta, T.R.; Gunnarsson, T.; Penn, I.; Ricci, D.R.; McDougall, I.; Marko, A.; Bourne, G.; Da Costa, L. A novel endovascular clip system for the treatment of intracranial aneurysms: Technology, concept, and initial experimental results: Laboratory investigation. *J. Neurosurg.* **2008**, *108*, 1230–1240. [CrossRef]
20. O'Kelly, C.; Rempel, J.L.; Diestro, J.D.B.; Marotta, T.R. Letter to the Editor: Pandemic (COVID-19) Proctoring for eCLIPs Neurointervention. *World Neurosurg.* **2020**, *142*, 575. [CrossRef]
21. Dholakia, R.; Sadasivan, C.; Fiorella, D.J.; Woo, H.H.; Lieber, B.B. Hemodynamics of flow diverters. *J. Biomech. Eng.* **2017**, *139*, BIO-16-1270. [CrossRef]
22. Yazdi, S.G.; Huetter, L.; Docherty, P.D.; Williamson, P.N.; Clucas, D.; Jermy, M.; Geoghegan, P.H. A Novel Fabrication Method for Compliant Silicone Phantoms of Arterial Geometry for Use in Particle Image Velocimetry of Haemodynamics. *Appl. Sci.* **2019**, *9*, 3811. [CrossRef]
23. Geoghegan, P.H.; Buchmann, N.A.; Spence, C.J.T.; Moore, S.; Jermy, M. Fabrication of rigid and flexible refractive-index-matched flow phantoms for flow visualisation and optical flow measurements. *Exp. Fluids* **2012**, *52*, 1331–1347. [CrossRef]
24. Johnston, I.D.; McCluskey, D.K.; Tan, C.K.L.; Tracey, M.C. Mechanical characterization of bulk Sylgard 184 for microfluidics and microengineering. *J. Micromech. Microeng.* **2014**, *24*, 35017. [CrossRef]
25. Yousif, M.Y.; Holdsworth, D.W.; Poepping, T.L. A blood-mimicking fluid for particle image velocimetry with silicone vascular models. *Exp. Fluids* **2011**, *50*, 769–774. [CrossRef]
26. Blanco, P.J.; Watanabe, S.M.; Dari, E.A.; Passos, M.A.R.F.; Feijóo, R.A. Blood flow distribution in an anatomically detailed arterial network model: Criteria and algorithms. *Biomech. Model. Mechanobiol.* **2014**, *13*, 1303–1330. [CrossRef] [PubMed]
27. Briganti, F.; Leone, G.; Marseglia, M.; Mariniello, G.; Caranci, F.; Brunetti, A.; Maiuri, F. Endovascular treatment of cerebral aneurysms using flow-diverter devices: A systematic review. *Neuroradiol. J.* **2015**, *28*, 365–375. [CrossRef]
28. Mut, F.; Raschi, M.; Scrivano, E.; Bleise, C.; Chudyk, J.; Ceratto, R.; Lylyk, P.; Cebral, J.R. Association between hemodynamic conditions and occlusion times after flow diversion in cerebral aneurysms. *J. Neurointerv. Surg.* **2015**, *7*, 286–290. [CrossRef]
29. Yu, C.H.; Matsumoto, K.; Shida, S.; Kim, D.J.; Ohta, M. A steady flow analsys on a cerebral aneurysm model with several stents for new stent design using PIV. *J. Mech. Sci. Technol.* **2012**, *26*, 1333–1340. [CrossRef]
30. Castro, M.A.; Putman, C.M.; Cebral, J.R. Computational fluid dynamics modeling of intracranial aneurysms: Effects of parent artery segmentation on intra-aneurysmal hemodynamics. *Am. J. Neuroradiol.* **2006**, *27*, 1703–1709.
31. Fiorella, D.; Arthur, A.S.; Chiacchierini, R.; Emery, E.; Molyneux, A.; Pierot, L. How safe and effective are existing treatments for wide-necked bifurcation aneurysms? Literature-based objective performance criteria for safety and effectiveness. *J. Neurointerv. Surg.* **2017**, *9*, 1197–1201. [CrossRef]
32. Kim, Y.H.; Xu, X.; Lee, J.S. The effect of stent porosity and strut shape on saccular aneurysm and its numerical analysis with lattice Boltzmann method. *Ann. Biomed. Eng.* **2010**, *38*, 2274–2292. [CrossRef] [PubMed]
33. Najjari, M.R.; Hinke, J.A.; Bulusu, K.V.; Plesniak, M.W. On the rheology of refractive-index-matched, non-Newtonian blood-analog fluids for PIV experiments. *Exp. Fluids* **2016**, *57*, 96. [CrossRef]
34. Frolov, S.V.; Sindeev, S.V.; Liepsch, D.; Balasso, A. Experimental and CFD flow studies in an intracranial aneurysm model with Newtonian and non-Newtonian fluids. *Technol. Health Care* **2016**, *24*, 317–333. [CrossRef] [PubMed]
35. Valencia, A.; Zarate, A.; Galvez, M.; Badilla, L. Non-Newtonian blood flow dynamics in a right internal carotid artery with a saccular aneurysm. *Int. J. Numer. Methods Fluids* **2006**, *50*, 751–764. [CrossRef]
36. Babiker, M.H.; Gonzalez, L.F.; Albuquerque, F.; Collins, D.; Elvikis, A.; Zwart, C.; Roszelle, B.; Frakes, D.H. An in vitro study of pulsatile fluid dynamics in intracranial aneurysm models treated with embolic coils and flow diverters. *IEEE Trans. Biomed. Eng.* **2012**, *60*, 1150–1159. [CrossRef] [PubMed]
37. Nair, P.; Chong, B.W.; Indahlastari, A.; Ryan, J.; Workman, C.; Babiker, M.H.; Farsani, H.Y.; Baccin, C.E.; Frakes, D. Hemodynamic characterization of geometric cerebral aneurysm templates treated with embolic coils. *J. Biomech. Eng.* **2016**, *138*, 021011. [CrossRef]
38. Nguyen, Y.N.; Kabinejadian, F.; Ismail, M.; Kong, W.K.F.; Tay, E.L.W.; Leo, H.L. Ex vivo assessment of bicuspidization repair in treating severe functional tricuspid regurgitation: A stereo-scopic PIV study. *Sci. Rep.* **2019**, *9*, 11504. [CrossRef]

39. Medero, R.; Ruedinger, K.; Rutkowski, D.; Johnson, K.; Roldán-Alzate, A. In Vitro Assessment of Flow Variability in an Intracranial Aneurysm Model Using 4D Flow MRI and Tomographic PIV. *Ann. Biomed. Eng.* **2020**, *48*, 2484–2493. [CrossRef]
40. Yamashita, S.; Isoda, H.; Hirano, M.; Takeda, H.; Inagawa, S.; Takehara, Y.; Alley, M.T.; Markl, M.; Pelc, N.J.; Sakahara, H. Visualization of hemodynamics in intracranial arteries using time-resolved three-dimensional phase-contrast MRI. *J. Magn. Reson. Imaging Off. J. Int. Soc. Magn. Reson. Med.* **2007**, *25*, 473–478. [CrossRef]

Publisher's Note: MDPI stays neutral with regard to jurisdictional claims in published maps and institutional affiliations.

applied sciences

MDPI

Article

Thermal Accelerometer Simulation by the R-Functions Method

Mikhail Basarab [1,*], Alain Giani [2] and Philippe Combette [2]

[1] Computer Science and Control Systems Department, Bauman Moscow State Technical University, 105005 Moscow, Russia

[2] Institut d'Electronique et des Systèmes (IES), Université de Montpellier, 34095 Montpellier, CEDEX 05, France; alain.giani@ies.univmontp2.fr (A.G.); philippe.combette@ies.univ-montp2.fr (P.C.)

* Correspondence: bmic@mail.ru; Tel.: +7-903-760-5499

Received: 1 November 2020; Accepted: 23 November 2020; Published: 25 November 2020

check for updates

Abstract: As well as many modern devices, thermal accelerometers (TAs) need a sophisticated mathematical simulation to find the ways for their performance optimization. In the paper, a novel approach for solving computational fluid dynamics (CFD) problems in the TA's cavity is proposed (MQ-RFM), which is based on the combined use of Rvachev's R-functions method (RFM) and the Galerkin technique with multiquadric (MQ) radial basis functions (RBFs). The semi-analytical RFM takes an intermediate position between traditional analytical approaches and numerical methods, such as the finite-element method (FEM), belonging to the family of the so-called meshless techniques which became popular in the last decades in solving various CFD problems in complex-shaped cavities. Mathematical simulation of TA by using the MQ-RFM was carried out with the purpose to simulate the temperature response of the device and to study and improve its performance. The results of numerical experiments were compared with well-known analytical and numerical benchmark solutions for the circular annulus geometry and it demonstrated the effectiveness of the MQ-RFM for solving the convective heat-transfer problem in the TA's cavity. The use of solution structures allows one to take a relatively small number of expansion terms to achieve an appropriate accuracy of the approximate solution satisfying at the same time the given boundary conditions exactly. The application of the MQ-RFM gives the possibility to obtain semi-analytical solutions to the diffusion-convection problems and to identify the main thermal characteristics of the TA, that allows one to improve the device performance.

Keywords: natural convection; Galerkin method; R-functions method; multiquadric RBF; thermal accelerometer; cylindrical annulus

1. Introduction

Many modern devices, in particular, sensors based on microelectromechanical systems (MEMS), need a sophisticated mathematical simulation to find the ways of their optimization. Among them are thermal accelerometers (TAs) [1] which principle of operation is based on the effect of fluid or gas convection in closed cavities under the influence of external forces of inertia. It is very important to estimate the scale factor and bias stabilities of TAs under an external thermal slope, and to test different types of cavity geometry (cylindrical, rectangular, hexagonal, etc.) to achieve the best performance of the device.

To solve the aforementioned problems, it is necessary to simulate diffusive and convective heat and mass transfer in arbitrarily shaped enclosures, which in turn is a problem of great importance due to a large number of its practical applications. Since analytical solutions exist for a very narrow class of simple-shaped domains only, a majority of modern computational fluid dynamics (CFD) techniques were developed which are based mainly on the use of the finite-difference and finite- element methods (FDM and FEM) [2–4], boundary-element methods [5], and spectral element methods [6]. Their main drawback is the cumbersome representation of grid solutions in complex- shaped domains, as well as the difficulties connected with the approximation of boundary conditions. For the function of pressure or for the vorticity function in dimensionless form, boundary conditions are not specified at all and require special approximations. In addition, sometimes it is difficult to interpret analytically the solutions obtained at a number of mesh knots and to evaluate errors.

Recently, a number of meshless (or mesh-free) techniques were proposed for solving convection-diffusion problems with arbitrarily shaped cavities [7–9]. They became popular due to their simplicity, flexibility, and independence from a complicated domain geometry. One of the most effective meshless techniques is the Rvachev's R-function method (RFM) [10]. It allows the given boundary conditions to be satisfied exactly at all boundary points by means of the appropriate transformation of the basic functions. With R-functions, it became possible to construct the functions with prescribed values and derivatives at specified locations. Furthermore, the constructed functions possess desired differential properties and may be assembled into a solution structure for the posed boundary value problems. The semi-analytical method of R-functions takes an intermediate position between traditional analytical approaches and numerical tools. Many practical problems in different areas of mathematical physics, in particular, heat conductivity [11], are being solved effectively by using RFM.

Application of the RFM in combination with the Galerkin technique for solving CFD problems in arbitrarily shaped domains was investigated earlier [12–14]. Different bases were applied in these works, both spectral (polynomial) and local (B-splines), and good results were achieved for fields evaluated in domains of simple geometry without localized inhomogeneities. In [15], an approach combining the RFM and the Petrov–Galerkin method with bases of algebraic polynomials was presented and applied to the simulation of thermal convection fields inside a closed rectangular cavity.

However, when applying the RFM, if a domain consists of two parts, for example, "inner" and "outer", which characteristic dimensions are not comparable, one needs to take either a large higher power of approximating polynomials (in spectral methods) that worsens computational stability, or a very small regular or irregular mesh width for bases of compactly supported functions. Therefore, in the latter case, the RFM loses its main benefits and can be practically considered as a conventional grid method such as FDM or FEM.

Here, a novel technique is proposed (MQ-RFM), based on the RFM with multiquadric (MQ) radial basis functions (RBFs) [16,17]. These functions are very popular as "building blocks" in different meshless methods intended for solving CFD problems [18,19]. Multiquadrics have a simple algebraic representation and, like algebraic or trigonometric polynomials, are spectral ones, i.e., not compactly supported. On the other hand, the basis of MQs can be simply formed on a usual regular mesh in the way similar to the bases of compactly supported bilinear or high-order splines. For the first time, the idea of using the RFM in combination with RBFs for solving linear boundary value problems for partial differential equations (PDEs) was proposed in [20], where irregular mesh was constructed on the base of the Voronoi diagrams.

In this work, the standard model of circular annulus geometry is studied which earlier was extensively investigated experimentally [21–23], analytically [24,25], and numerically [26–33]. This model is the basic one for studying the TA behavior and performance [34–36]. The distance between the heater and detectors was found providing an optimum sensitivity of the sensor.

The results of numerical experiments confirm the effectiveness of RFM applied for solving the convective heat-transfer problem in the TA's cavity and also demonstrate good accordance with the results obtained experimentally, analytically, and numerically with the help of FEM simulation [34,36]. The use of RFM solution structures and MQ-RBFs defined on a simple regular mesh allows one to take a relatively small number of expansion terms to achieve an appropriate accuracy of the approximate solution satisfying the given boundary conditions exactly. The application of the MQ-RFM gives a possibility to obtain the semi-analytical solutions to the diffusion-convection problems and identify main thermal characteristics of the TA of more complicated geometry [37], that allows one to improve the device performance.

2. Thermal Accelerometer and Its Principle of Operation

The principle of operation of a TA is based on the effect of fluid or gas convection in a closed cavity under the influence of external forces of inertia. The device (Figure 1) includes a heating element H that creates around itself a symmetrical thermal field. The thermal sensors are located on opposite sides of this element.

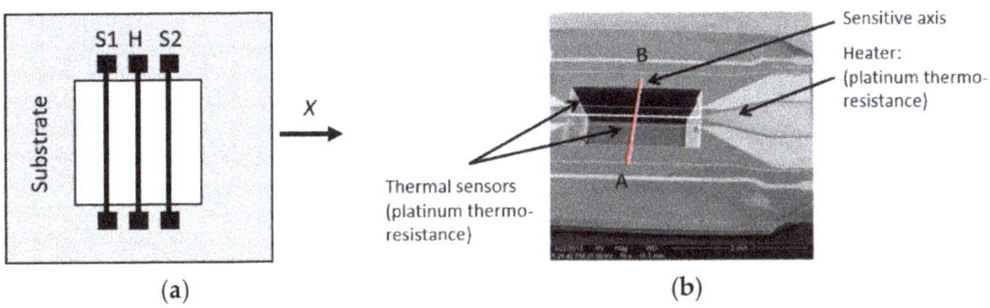

Figure 1. The general scheme (**a**) and microstructure design (**b**) of a thermal accelerometers (TA).

In the absence of an external acceleration along the sensitive axis X, the system is balanced and the heat detectors generate the same signal (Figure 2a). In presence of an external acceleration, the temperature profile changes, which results in the temperature difference dT between the sensors S1 and S2, depending on the amplitude of the acceleration (Figure 2b). Nonzero temperature difference between the heat sensors converts the input impedance to the output electrical signal.

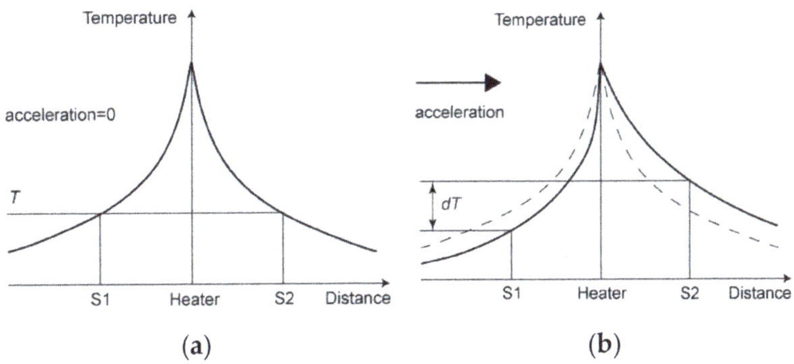

Figure 2. The principle of operation of a TA. (**a**) without acceleration, (**b**) with acceleration.

This type of accelerometer has several important advantages over conventional sensors for acceleration based on microelectromechanical systems (MEMS). In particular, due to the absence of moving mechanical parts, convective accelerometers have high reliability, low cost of production, as well as the ability to withstand and measure high loads caused, for example, by shock action.

To find the optimal parameters of the device: the type of gas and its pressure, the size and geometry of the cavity, the material and the size of the heating element and heat sensors, etc., it is necessary to solve the boundary value problem described by a system of the Navier–Stokes differential equations. This problem can be solved analytically only for a limited class of simple regions (cylinder, sphere), and in the general case, a numerical simulation is required.

3. Statement of the Problem and Governing Equations

Usually, the temperature distribution inside the 3D cavity of the device is almost uniform in the direction along the heater and sensors. Thus we need to solve the 2D heat-transfer problem in the middle cross-section of the cavity. One of the most important models of the TA is the circular annulus model (Figure 3a) [34,36] because for such a domain the analytical solution to the heat diffusion problem is known as well as the asymptotic solution of the diffusion-convection problem [24,25].

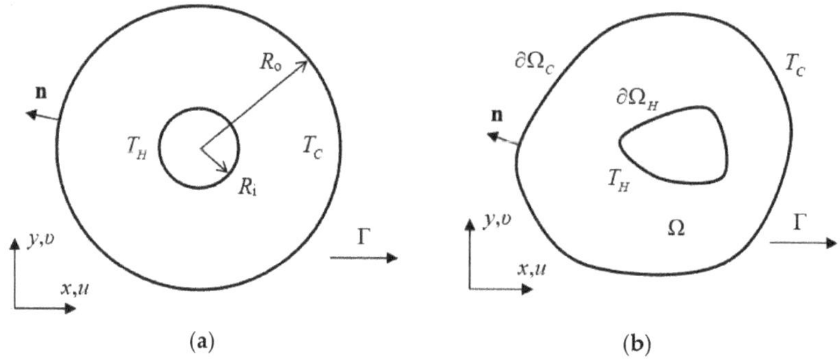

Figure 3. Convection-diffusion problem in a circular annulus (**a**) and geometry of a common boundary-value problem (**b**).

We consider the Boussinesq approximation with the assumption of the steady state incompressible flow inside an arbitrary 2D closed cavity $\Omega \subset \mathbf{R}^2$ (Figure 3b), when all constants, except density, do not depend on temperature. Without loss of generality, we take the domain with boundary composed of two parts: $\partial\Omega_C$, $\partial\Omega_H$, with given constant temperatures T_C and T_H on each of them respectively. Let external acceleration Γ be applied towards the positive direction of sensitivity axis OX.

The dimensionless governing equations for the two-dimensional convective heat and mass transfer inside Ω have the following form:

$$\begin{aligned} &\tfrac{1}{\mathrm{Pr}}\nabla^2\theta - \left(U\tfrac{\partial\theta}{\partial X} + V\tfrac{\partial\theta}{\partial Y}\right) = 0; \\ &\nabla^2\zeta - \left(U\tfrac{\partial\zeta}{\partial X} + V\tfrac{\partial\zeta}{\partial Y}\right) = -\mathrm{Gr}\tfrac{\partial\theta}{\partial Y}; \\ &\nabla^2\psi = -\zeta;\ U = \tfrac{\partial\psi}{\partial Y},\ V = -\tfrac{\partial\psi}{\partial X}, \end{aligned} \tag{1}$$

where $\theta = (T - T_C)/(T_H - T_C)$ is the dimensionless temperature; T is the temperature; ψ is the stream function; ζ is the vorticity; $X = x/L$, $Y = y/L$ are dimensionless coordinates; L is the specific dimension; $U = \mathrm{u}L/\nu$; $V = \mathrm{v}L/\nu$ are dimensionless components of velocity; u, v are horizontal and vertical components of velocity; ν is the kinematic viscosity; $\mathrm{Gr} = \Gamma\beta(T_H - T_C)L^3/\nu^2$ is the Grashof number; Γ is the acceleration; β is the temperature volume expansion coefficient; $\mathrm{Pr} = c_p\mu/\lambda$ is the Prandtl number; μ is the dynamic viscosity; λ is the heat conductivity; c_p is the specific heat at constant pressure.

Therefore, on $\partial\Omega$ the following dimensionless boundary conditions are given:

$$\begin{aligned} &\partial\Omega_C:\ \theta = 0; \\ &\partial\Omega_H:\ \theta = 1; \\ &\partial\Omega:\ \psi = \tfrac{\partial\psi}{\partial \mathbf{n}} = U = V = 0, \end{aligned} \tag{2}$$

where $\mathbf{n}$ is the outward normal to boundary $\partial\Omega$.

Boundary conditions for the vorticity function ζ are not defined explicitly and are usually approximated by its Taylor series expansion in the neighborhood of the boundary.

4. Generalized Numerical Procedure

First of all, before applying common variational techniques considered in [38,39], we should pass from inhomogeneous boundary conditions (2) to homogeneous ones by introducing the new function:

$$\widetilde{\theta} = \theta - \Phi, \tag{3}$$

where Φ is a continuous function which is equal to unity on the part $\partial\Omega_H$ and vanishes on $\partial\Omega_C$. Function Φ can be constructed in the way similar to that for the traditional Lagrange interpolation polynomial but here instead of interpolation points we take corresponding segments of the boundary. This technique also is called the transfinite interpolation [40].

Hereinafter, to avoid cumbersome designations, we shall denote:

$$\theta \equiv \widetilde{\theta}.$$

Thus, we get:

$$\begin{gathered} \tfrac{1}{\Pr}\nabla^2\theta - \left(U\tfrac{\partial\theta}{\partial X} + V\tfrac{\partial\theta}{\partial Y}\right) = -\tfrac{1}{\Pr}\nabla^2\Phi + U\tfrac{\partial\Phi}{\partial X} + V\tfrac{\partial\Phi}{\partial Y}; \\ \nabla^2\zeta - \left(U\tfrac{\partial\zeta}{\partial X} + V\tfrac{\partial\zeta}{\partial Y}\right) = -\mathrm{Gr}\cdot\left(\tfrac{\partial\theta}{\partial Y} + \tfrac{\partial\Phi}{\partial Y}\right); \\ \nabla^2\psi = -\zeta;\ U = \tfrac{\partial\psi}{\partial Y},\ V = -\tfrac{\partial\psi}{\partial X}; \end{gathered} \tag{4}$$

$$\partial\Omega:\ \theta = \psi = \frac{\partial\psi}{\partial\mathbf{n}} = U = V = 0. \tag{5}$$

Equations of System (4) can be solved iteratively:

$$\begin{gathered} \tfrac{1}{\Pr}\nabla^2\theta^{(k)} - \left(U^{(k-1)}\tfrac{\partial\theta^{(k)}}{\partial X} + V^{(k-1)}\tfrac{\partial\theta^{(k)}}{\partial Y}\right) = -\tfrac{1}{\Pr}\nabla^2\Phi + U^{(k-1)}\tfrac{\partial\Phi}{\partial X} + V^{(k-1)}\tfrac{\partial\Phi}{\partial Y}; \\ \nabla^2\zeta^{(k)} - \left(U^{(k-1)}\tfrac{\partial\zeta^{(k)}}{\partial X} + V^{(k-1)}\tfrac{\partial\zeta^{(k)}}{\partial Y}\right) = -\mathrm{Gr}\left(\tfrac{\partial\theta^{(k)}}{\partial Y} + \tfrac{\partial\Phi}{\partial Y}\right); \\ \nabla^2\psi^{(k)} = -\zeta^{(k)};\ U^{(k)} = \tfrac{\partial\psi^{(k)}}{\partial Y},\ V^{(k)} = -\tfrac{\partial\psi^{(k)}}{\partial X};\ (k = 1, 2, \ldots). \end{gathered} \tag{6}$$

Initial approximations $\zeta^{(0)},\ \psi^{(0)},\ U^{(0)},\ V^{(0)}$ must be given. If $\zeta^{(0)} = \psi^{(0)} = U^{(0)} = V^{(0)} = 0$, then after the first step ($k = 1$) we obtain the solution to the stationary heat diffusion problem without convection.

The iterative process (6) stops when some convergence conditions are fulfilled, for example,

$$\frac{\left\|\theta^{(k+1)} - \theta^{(k)}\right\|_{\mathbf{L}_2(\Omega)}}{\left\|\theta^{(k+1)}\right\|_{\mathbf{L}_2(\Omega)}} \le \varepsilon << 1. \tag{7}$$

In a number of variational and projection techniques [6,38,39], approximations to $\theta^{(k)}$ at each iterative step are found in the form of truncated generalized Fourier series with respect to functions of some basis $\{f_n\}_{n=0}^{N}$:

$$\theta^{(k)} \approx \sum_{n=0}^{N} c_n^{(k)} f_n, \tag{8}$$

where $\mathbf{c}^{(k)} = (c_0^{(k)}, c_1^{(k)}, \ldots, c_N^{(k)})$ are undefined coefficients.

In addition to their differentiability, all functions f_n must vanish on the boundary, i.e.,

$$\partial\Omega:\ f_n = 0,\ n = 0, \ldots, N. \tag{9}$$

From Equation (9) it follows that approximation (8) at any iterative step will also satisfy the homogeneous boundary condition (5) for θ exactly.

Leonid Kantorovich [38] proposed the following technique for constructing a functional basis satisfying Equation (9). First we should take an arbitrary basis system of functions $\{\chi_n\}_{n=0}^{N}$. There may be algebraic or trigonometric polynomials, splines, etc. Then a function $\omega(X,Y)$ is constructed such that $\omega > 0$ inside Ω; $\omega < 0$ outside $\Omega \cup \partial\Omega$; $\omega = 0$ and $|\nabla\omega|_2 \neq 0$ on $\partial\Omega$. For simple domains, for example circular, these functions are trivial, but the common approach to obtaining analogous expressions for arbitrary domains was developed by V. Rvachev [10]. This approach (the R-functions method—RFM) will be considered in the next section.

Then we take the new system of functions

$$f_n \equiv \omega\chi_n,\ n = 0,\ldots,N, \tag{10}$$

which, due to the properties of the function ω, satisfy the conditions (9).

Taking into account Equation (9), Equation (8) can be written as:

$$\theta^{(k)} \approx \sum_{n=0}^{N} c_n^{(k)} f_n = \omega \sum_{n=0}^{N} c_n^{(k)} \chi_n. \tag{11}$$

To strictly obey conditions (5) for the stream function, we must take another basis $\left\{\omega^2\chi_n\right\}_{n=0}^{N}$ with undefined coefficients $\mathbf{e}^{(k)} = (e_0^{(k)}, e_1^{(k)}, \ldots, e_N^{(k)})$:

$$\psi^{(k)} \approx \sum_{n=0}^{N} e_n^{(k)} \omega f_n = \omega^2 \sum_{n=0}^{N} e_n^{(k)} \chi_n. \tag{12}$$

Then the velocity components will be expressed as

$$U^{(k)} = \sum_{n=0}^{N} e_n^{(k)} \frac{\partial(\omega f_n)}{\partial Y},\ V^{(k)} = -\sum_{n=0}^{N} e_n^{(k)} \frac{\partial(\omega f_n)}{\partial X}. \tag{13}$$

Substituting Equation (12) into the left-hand side of the third equation of system (4), we obtain a vorticity function expansion:

$$\zeta^{(k)} \approx \sum_{n=0}^{N} d_n^{(k)} \nabla^2(\omega f_n). \tag{14}$$

The latter expression may be considered as a representation of the vorticity function with respect to the set of basic functions $\nabla^2(\omega f_n)$ with a set of unknown coefficients $\mathbf{d}^{(k)} = (d_0^{(k)}, d_1^{(k)}, \ldots, d_N^{(k)})$:

$$d_n^{(k)} = -e_n^{(k)},\ n = 0,\ldots,N. \tag{15}$$

Thus, it is necessary to solve recurrently only two equations of System (6) to find coefficients $c_n^{(k)}$ and $d_n^{(k)}$. It should also be noted here that Equation (14) provides us an expression for the vorticity functions without the necessity to approximate additionally its boundary values.

At each iteration, the realization of various numerical techniques requires different procedures of reducing an original problem to a corresponding system of linear algebraic equations (SLAE) with respect to unknown vectors $\mathbf{c}^{(k)}$ and $\mathbf{d}^{(k)}$. Here we propose to use an approach on the base of the Petrov–Galerkin procedure [6].

At the first step, we find the components $c_n^{(k)}$ by the Galerkin technique. Substituting Equation (11) into the first equation of (6), we obtain the residual:

$$\delta_\theta^{(k)} = \sum_{n=0}^{N} c_n^{(k)} \left(\frac{1}{\mathrm{Pr}} \nabla^2 f_n - U^{(k-1)} \frac{\partial f_n}{\partial X} - V^{(k-1)} \frac{\partial f_n}{\partial Y} \right) - F_\theta^{(k-1)}. \tag{16}$$

Here the right-hand part is

$$F_{\theta}^{(k-1)} = -\left(\frac{1}{\Pr}\nabla^2\Phi - U^{(k-1)}\frac{\partial\Phi}{\partial X} - V^{(k-1)}\frac{\partial\Phi}{\partial Y}\right).$$

One must choose the set of coefficients $c_n^{(k)}$ minimizing residual (16). In the Galerkin scheme, the orthogonality of δ_θ to all functions of the system $\{f_n\}_{n=0}^N$ in space $\mathbf{L}_2(\Omega)$ is required, i.e.,

$$\int\limits_{\Omega} \delta_\theta^{(k)} f_m \mathrm{d}\sigma = 0,\ m = 0, \ldots, N, \tag{17}$$

where $\mathrm{d}\sigma \equiv \mathrm{d}X\mathrm{d}Y$.

Substituting Equation (16) in Equation (17), we get the SLAE with respect to elements of the vector $\mathbf{c}^{(k)}$:

$$(\mathbf{A} + \breve{\mathbf{A}}^{(k-1)})\mathbf{c}^{(k)} = \mathbf{b} + \breve{\mathbf{b}}^{(k-1)}, \tag{18}$$

where components of matrices $\mathbf{A}, \breve{\mathbf{A}}^{(k-1)}$ and vectors $\mathbf{b}, \breve{\mathbf{b}}^{(k-1)}$ are determined respectively as:

$$\begin{gathered}
a_{m,n} = \tfrac{1}{\Pr}\int\limits_{\Omega} f_m \nabla^2 f_n \mathrm{d}\sigma = -\tfrac{1}{\Pr}\int\limits_{\Omega} \nabla f_m \nabla f_n \mathrm{d}\sigma, \\
\breve{a}_{m,n}^{(k-1)} = -\int\limits_{\Omega} f_m\left(U^{(k-1)}\tfrac{\partial f_n}{\partial X} + V^{(k-1)}\tfrac{\partial f_n}{\partial Y}\right)\mathrm{d}\sigma; \\
b_m = -\tfrac{1}{\Pr}\int\limits_{\Omega} f_m \nabla^2\Phi\, dXdY = \tfrac{1}{\Pr}\int\limits_{\Omega} \nabla f_m \nabla\Phi\, \mathrm{d}\sigma, \\
\breve{b}_m^{(k-1)} = -\int\limits_{\Omega} f_m\left(U^{(k-1)}\tfrac{\partial\Phi}{\partial X} + V^{(k-1)}\tfrac{\partial\Phi}{\partial Y}\right)\mathrm{d}\sigma; \\
m, n = 0, \ldots, N.
\end{gathered} \tag{19}$$

Some optimization of Equation (19) can be done to decrease the time for computations.

Then we pass to finding coefficients of expansion (14). Analogously to the previous case, we write the corresponding residual for the second equation of the System (6) as:

$$\delta_\zeta^{(k)} = \sum_{n=1}^{N} d_n^{(k)}\left(\nabla^2\nabla^2(\omega f_n) - U^{(k-1)}\frac{\partial}{\partial X}\nabla^2(\omega f_n) - V^{(k-1)}\frac{\partial}{\partial Y}\nabla^2(\omega f_n)\right) - F_\zeta^{(k)}, \tag{20}$$

where

$$F_\zeta^{(k)} = -\mathrm{Gr}\cdot\left(\frac{\partial\theta^{(k)}}{\partial Y} + \frac{\partial\Phi}{\partial Y}\right).$$

To find coefficients $d_n^{(k)}$, we make an orthogonal projection of the residual (20) to all functions of the basis $\{f_n\}_{n=0}^N$ that is the main idea of the Petrov–Galerkin method [6]:

$$\int\limits_{\Omega} f_m \delta_\zeta^{(k)} \mathrm{d}\sigma = 0,\ m = 0, \ldots, N. \tag{21}$$

Finally we obtain the following SLAE for $\mathbf{d}^{(k)}$:

$$(\mathbf{W} + \breve{\mathbf{W}}^{(k-1)})\mathbf{d}^{(k)} = \mathbf{z} + \breve{\mathbf{z}}^{(k)}, \tag{22}$$

where

$$\begin{aligned}
& w_{m,n} = \int_{\Omega} f_m \nabla^2 \nabla^2 (\omega f_n)\, d\sigma; \\
& \overset{\smile}{w}{}^{(k-1)}_{m,n} = -\int_{\Omega} f_m \Big(U^{(k-1)} \tfrac{\partial}{\partial X} \nabla^2 (\omega f_n) + V^{(k-1)} \tfrac{\partial}{\partial Y} \nabla^2 (\omega f_n) \Big) d\sigma; \\
& z_m = -\mathrm{Gr} \int_{\Omega} f_m \tfrac{\partial \Phi}{\partial Y} d\sigma; \\
& \overset{\smile}{z}{}^{(k)}_{m} = -\mathrm{Gr} \int_{\Omega} f_m \tfrac{\partial \theta^{(k)}}{\partial Y} d\sigma, \ m, n = 1, \ldots, N.
\end{aligned} \tag{23}$$

Integrals in Equations (19) and (23) can be evaluated numerically, and differential operators are approximated with their finite-difference analogs.

5. R-Functions Method and Transfinite Interpolation

The RFM [10] allows all prescribed boundary conditions to be satisfied exactly at all boundary points. The R-functions are real-valued functions that behave as continuous analogs of logical Boolean functions. With R-functions, it became possible to construct functions with prescribed values and derivatives at specified locations. Furthermore, the constructed functions possess desired differential properties and may be assembled into a solution structure that is guaranteed to contain solutions to the posed boundary value problems. For example, the homogeneous Dirichlet conditions may be satisfied exactly by representing the solution as the product of two functions: (i) a real-valued R-function that takes on zero values at the boundary points; and (ii) an unknown function that allows one to satisfy (exactly or approximately) the differential equation of the problem.

A function $f(x, y)$ is called an *R-function* if its sign is completely determined by the signs (but not magnitudes) of its arguments.

The most popular system of R-functions is the system with R-operations (R-conjunction, R-disjunction, and R-negation) defined in the following way:

$$\begin{aligned}
& x \wedge y \equiv x + y - \sqrt{x^2 + y^2}, \\
& x \vee y \equiv x + y + \sqrt{x^2 + y^2}, \\
& \overline{x} \equiv -x.
\end{aligned} \tag{24}$$

The above R-functions correspond to the Boolean logic functions $\wedge$, $\vee$, $\neg$ in a piecewise sense and allow constructing normalized implicit functions for complex-shaped geometric objects. Let the geometric domain $\Omega = B(\Omega_C, \Omega_H)$ be constructed as a Boolean (union and intersection) combination of primitive regions Ω_C, Ω_H, defined by real-valued functional inequalities $\omega_C(x, y) > 0$, $\omega_H(x, y) > 0$ respectively. If f is an R-function corresponding to the Boolean function B, then the implicit function of the resulting geometric domain is immediately given by $\Omega = [f(\omega_C, \omega_H) > 0]$. The function $f(\omega_C, \omega_H)$ is negative outside of Ω and the equation $f(\omega_C, \omega_H) = 0$ defines the boundary $\partial\Omega$ of the domain Ω.

It is known that the equation of the boundary $\partial\Omega$ $(f = 0)$ is called normal if the value of $f(x, y)$ is equal to the Euclidean distance from the point (x, y) to the boundary $\partial\Omega$. Similarly a function f that coincides with the normal function only on the boundary $\partial\Omega$ is called normalized and has a property that:

$$\left. \frac{\partial f}{\partial \mathbf{n}} \right|_{\partial\Omega} = -1. \tag{25}$$

If both implicit functions, ω_C, ω_H, are normalized on the boundaries Ω_C, Ω_H respectively then all of the R-functions above preserve this property, and the function $f(\omega_C, \omega_H)$ is normalized on the whole boundary $\partial\Omega$.

With the help of R-functions, it is possible to make the transfinite interpolation, i.e., to construct a continuous expression satisfying different conditions on boundary parts $\partial\Omega_C$, $\partial\Omega_H$. Here, the "bonding" operation for boundary conditions must be used. For example, the function:

$$\Phi = \frac{\theta_H \omega_C + \theta_C \omega_H}{\omega_C + \omega_H} \tag{26}$$

satisfies the conditions

$$\Phi|_{\partial\Omega_C} = \theta_C, \ \Phi|_{\partial\Omega_H} = \theta_H. \tag{27}$$

The latter expression is a generalized analog of the Lagrange interpolation formula and allows us to pass from inhomogeneous boundary conditions to homogeneous ones.

The detailed justification of the RFM with Galerkin technique in connection with solving the natural convection problem in enclosure regions is presented in [12,41,42]. In particular, the natural convection in presence of local heat is investigated in [12]. This justification is based on variational principles [43] and is appropriate for the wide class of bases composed of both spectral and compactly supported functions.

It is convenient to take normalized functions due to the fact that with them it is easier to evaluate some important differential characteristics of the solution, in particular, the Nusselt number, Nu. The local Nusselt number at a point of boundary $\partial\Omega$ is expressed as:

$$\mathrm{Nu}|_{\partial\Omega} = -\left.\frac{\partial\theta}{\partial\mathbf{n}}\right|_{\partial\Omega}. \tag{28}$$

Taking into account Equations (3) and (11) and properties of the normalized function ω, we can write that:

$$\mathrm{Nu}|_{\partial\Omega} = -\frac{\partial}{\partial\mathbf{n}}\left.\left(\omega\sum_{n=0}^{N} c_n^{(k)}\chi_n + \Phi\right)\right|_{\partial\Omega} = -\left.\left(\frac{\partial\omega}{\partial\mathbf{n}}\sum_{n=0}^{N} c_n^{(k)}\chi_n + \omega\sum_{n=0}^{N} c_n^{(k)}\frac{\partial\chi_n}{\partial\mathbf{n}} + \frac{\partial\Phi}{\partial\mathbf{n}}\right)\right|_{\partial\Omega} = \left.\left(\sum_{n=0}^{N} c_n^{(k)}\chi_n - \frac{\partial\Phi}{\partial\mathbf{n}}\right)\right|_{\partial\Omega}. \tag{29}$$

Substituting Equation (26) into Equation (29) and taking into account properties of the normalized functions ω_C, ω_H, we get:

$$\left.\frac{\partial\Phi}{\partial\mathbf{n}}\right|_{\partial\Omega} = \frac{\partial}{\partial\mathbf{n}}\left.\left(\frac{\theta_H\omega_C + \theta_C\omega_H}{\omega_C + \omega_H}\right)\right|_{\partial\Omega} = \frac{1}{\omega_C + \omega_H}\left.\left[\left(\theta_H\frac{\partial\omega_C}{\partial\mathbf{n}} + \theta_C\frac{\partial\omega_H}{\partial\mathbf{n}}\right) - \Phi\left(\frac{\partial\omega_C}{\partial\mathbf{n}} + \frac{\partial\omega_H}{\partial\mathbf{n}}\right)\right]\right|_{\partial\Omega}.$$

Therefore on each part of the boundary $\partial\Omega$ we have the simple expressions:

$$\left.\frac{\partial\Phi}{\partial\mathbf{n}}\right|_{\partial\Omega_C} = (\theta_C - \theta_H)\left.\frac{1}{\omega_H}\right|_{\partial\Omega_C}, \ \left.\frac{\partial\Phi}{\partial\mathbf{n}}\right|_{\partial\Omega_H} = (\theta_H - \theta_C)\left.\frac{1}{\omega_C}\right|_{\partial\Omega_H}, \tag{30}$$

and

$$\mathrm{Nu}|_{\partial\Omega_C} = \sum_{n=0}^{N} c_n^{(k)}\chi_n|_{\partial\Omega_C} + (\theta_H - \theta_C)\left.\frac{1}{\omega_H}\right|_{\partial\Omega_C}, \ \mathrm{Nu}|_{\partial\Omega_H} = \sum_{n=0}^{N} c_n^{(k)}\chi_n|_{\partial\Omega_C} + (\theta_C - \theta_H)\left.\frac{1}{\omega_C}\right|_{\partial\Omega_H}. \tag{31}$$

Average Nusselt numbers on boundaries $\partial\Omega_C$, $\partial\Omega_H$ are found by integration of local Nusselt numbers along corresponding contours:

$$\left.\overline{\mathrm{Nu}}\right|_{\partial\Omega_C} = \frac{1}{|\partial\Omega_C|}\int_{\partial\Omega_C}\mathrm{Nu}ds, \ \left.\overline{\mathrm{Nu}}\right|_{\partial\Omega_H} = \frac{1}{|\partial\Omega_H|}\int_{\partial\Omega_H}\mathrm{Nu}ds. \tag{32}$$

6. Numerical Experiment

Consider the model boundary value problem in circular annulus (Figure 3a) with boundary conditions:

$$\begin{aligned} X^2+Y^2=R_i:\ \theta=1,\ \psi=\frac{\partial\psi}{\partial\mathbf{n}}=U=V=0; \\ X^2+Y^2=R_o:\ \theta=\psi=\frac{\partial\psi}{\partial\mathbf{n}}=U=V=0. \end{aligned} \tag{33}$$

The common way to obtain a function describing the boundary of the domain shown in Figure 3b is the use of R-conjunction operation:

$$\omega=\omega_C\wedge\omega_H=\omega_C+\omega_H-\sqrt{\omega_C^2+\omega_H^2}. \tag{34}$$

For the circular annulus, expressions for "hot" and "cold" parts of the boundary can be written as follows:

$$\omega_H(X,Y)=-\frac{1}{2R_\mathrm{i}}\left(R_\mathrm{i}^2-X^2-Y^2\right),\ \omega_C(X,Y)=\frac{1}{2R_\mathrm{o}}\left(R_\mathrm{o}^2-X^2-Y^2\right). \tag{35}$$

Here, the normalizing terms $1/(2R_{\mathrm{i}/o})$ are taken to provide the normalized functions ω_C,ω_H whose values, as well as values of ω, are close to the distances to corresponding boundaries at points located in their vicinity.

At first glance, for such a simple domain as shown in Figure 3a, by analogy with the approach proposed in [38,39], one would take instead of Equation (34) another expression:

$$\omega=\omega_C\omega_H=-\left(R_\mathrm{i}^2-X^2-Y^2\right)\left(R_\mathrm{o}^2-X^2-Y^2\right). \tag{36}$$

However, this way of constructing the function ω is restricted by the narrow class of domains (eccentric circular annuli, elliptic annuli, etc.) and it is inappropriate, for example, for annular domains with more complicated inner boundaries (triangular, rectangular, et al. [44,45]). Additionally, due to the fact that the function (36) is not normalized, namely,

$$\left.\frac{\partial\omega}{\partial\mathbf{n}}\right|_{\partial\Omega_C}=\omega_H|_{\partial\Omega_C},\ \left.\frac{\partial\omega}{\partial\mathbf{n}}\right|_{\partial\Omega_H}=\omega_C|_{\partial\Omega_H},$$

we shall obtain more complicated formulae for the local Nusselt numbers than Equation (31) while the application of functions (35) yields the following expressions:

$$\begin{aligned} \mathrm{Nu}|_{\partial\Omega_C}&=\sum_{n=0}^{N}c_n^{(k)}\chi_n|_{\partial\Omega_C}+2R_\mathrm{o}\frac{\theta_H-\theta_C}{R_\mathrm{o}^2-R_\mathrm{i}^2},\ \mathrm{Nu}|_{\partial\Omega_H}=\sum_{n=0}^{N}c_n^{(k)}\chi_n|_{\partial\Omega_H}+2R_\mathrm{i}\frac{\theta_C-\theta_H}{R_\mathrm{o}^2-R_\mathrm{i}^2}; \\ \overline{\mathrm{Nu}}\Big|_{\partial\Omega_C}&=\sum_{n=0}^{N}c_n^{(k)}\int_{\partial\Omega_C}\chi_n\mathrm{d}s+2R_\mathrm{o}\frac{\theta_H-\theta_C}{R_\mathrm{o}^2-R_\mathrm{i}^2},\ \overline{\mathrm{Nu}}\Big|_{\partial\Omega_H}=\sum_{n=0}^{N}c_n^{(k)}\int_{\partial\Omega_H}\chi_n\mathrm{d}s+2R_\mathrm{i}\frac{\theta_C-\theta_H}{R_\mathrm{o}^2-R_\mathrm{i}^2}. \end{aligned}$$

With the help of the transfinite interpolation (26), we pass to the problem with homogeneous boundary conditions:

$$\Phi=\frac{\omega_C}{\omega_H+\omega_C}. \tag{37}$$

In this work, we take the basis of multiquadric RBF (MQ-RBF) which is often used in various meshless techniques [18,19]:

$$\chi_n(X,Y)=\sqrt{\left(X-X_{i(n)}\right)^2+\left(Y-Y_{j(n)}\right)^2+\alpha^2}, \tag{38}$$

where $\left(X_{i(n)}Y_{j(n)}\right)$ are the centers of functions χ_n and α is their shape parameter. For simplicity we take $\left(X_{i(n)}Y_{j(n)}\right)$ as knots of a regular two-dimensional square mesh with widths $h_X=h_Y=\alpha$.

Instead of the latter representation we may construct 2D MQ-RBF by means of the tensor product:

$$\chi_n(X,Y) = \sqrt{\left(X - X_{i(n)}\right)^2 + \alpha^2} \cdot \sqrt{\left(Y - Y_{j(n)}\right)^2 + \alpha^2}. \tag{39}$$

We used both formulae, (38) and (39), and numerical experiments demonstrated that from the point of view of computation costs and accuracy they are almost equivalent.

To define the appropriate parameters of our technique, we tested it on the problem of the stationary heat diffusion problem in the annulus cavity, whose stationary axisymmetric analytical solution is:

$$T(r) = T_C + (T_H - T_C)\left(1 - \frac{\ln\left(R_{\mathrm{i}}^{-1}\right)}{\ln\left(R_{\mathrm{o}}R_{\mathrm{i}}^{-1}\right)}\right). \tag{40}$$

Figure 4 illustrates temperature isolines for stationary heat transfer problem in circular annulus with different ratios between inner and outer radii, which are in good accordance with the solution (40). Only 36 basic functions were taken with centers in knots of the square regular mesh. Numerical integration was realized by the two-dimensional method of trapezoids on the regular rectangular mesh of 32 × 32 nodes.

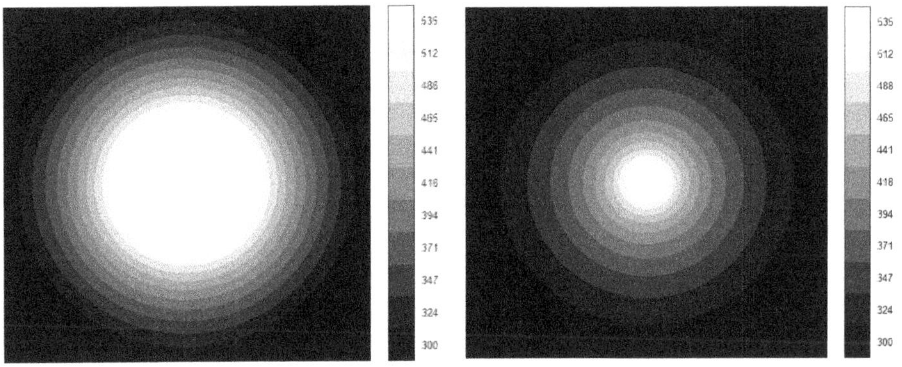

(a) (b)

Figure 4. Temperature isolines for different ratios: $R_i = R_o/2.6$ (**a**); $R_i = R_o/10$ (**b**).

The geometrical and thermophysical parameters were as follows:

$$R_o = 1.5 \text{ mm}; \; T_H = 535 \text{ K}, \; T_C = 300 \text{ K};$$

$$\rho = 0.54 \text{ kg/m}^3, \; c_p = 1000 \text{ J/(kg} \cdot K), \; \nu = 3.2 \cdot 10^{-5} \text{ m}^2/\text{s}, \; \beta = 1.5 \cdot 10^{-3} \text{ K}^{-1}.$$

Relative errors between approximate and exact solutions in $\mathbf{L}_2$-norm were $\varepsilon \approx 4 \cdot 10^{-4}$ for $R_i = R_o/2.6$ and $\varepsilon \approx 7 \cdot 10^{-3}$ for $R_i = R_o/10$. Table 1 presents the relative error as a function of the number of mesh nodes. The same resuls are shown in Figure 5 in logarithmic scale.

Table 1. Relative errors between approximate and analytical solutions as a function of the number of mesh nodes along each direction ($R_i = R_o/10$).

Number of Nodes	2	4	6	8	10	12
Relative error	0.019	0.012	7.3×10^{-3}	4.3×10^{-3}	2.6×10^{-3}	1.5×10^{-3}

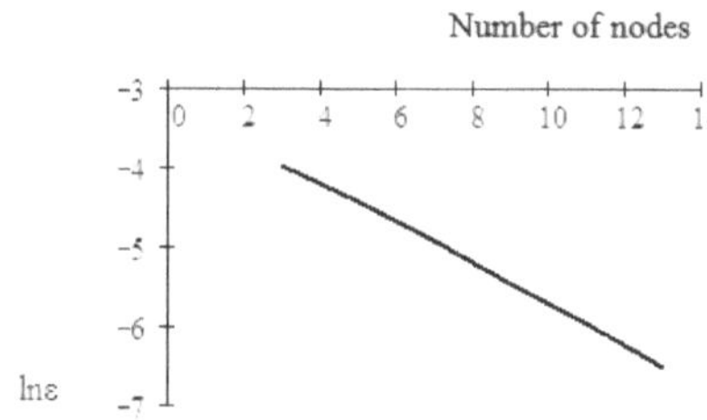

Figure 5. Logarithmic plot of the relative error between approximate and analytical solutions as a function of the number of mesh nodes along each direction ($R_i = R_o/10$).

Then we studied the stationary convective-diffusive heat transfer. For this set of parameters, we have $\mathrm{Pr} \approx 0.67$. In the case of a circular annulus, the characteristic length L in the expression for the Grashof number is equal to R_i, i.e., $\mathrm{Gr} = \Gamma\beta(T_H - T_C)R_\mathrm{i}^3/\nu^2$.

To increase the accuracy, instead of an expression (37) for "bonding" function Φ, we can take immediately the stationary heat diffusion distribution (40):

$$\Phi(X,Y) = T_C + (T_H - T_C)\left(1 - \frac{\ln\left(R_\mathrm{i}^{-1}\sqrt{X^2+Y^2}\right)}{\ln\left(R_\mathrm{o}R_\mathrm{i}^{-1}\right)}\right).$$

Figures 6 and 7 illustrate temperature and streamfunction isolines for both configurations respectively. Temperature distributions along the symmetry axis OX are shown in Figure 8.

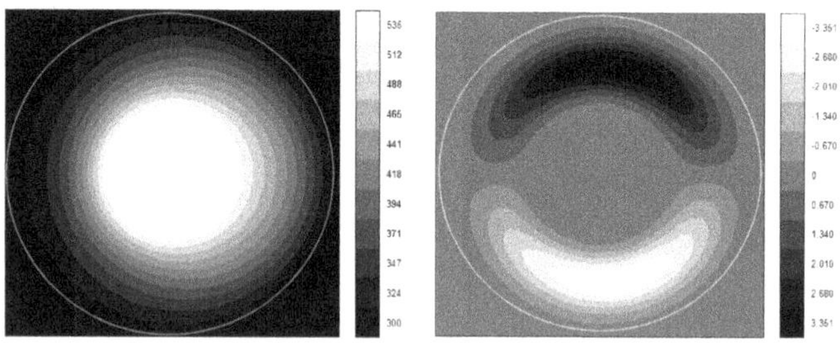

(a) (b)

Figure 6. Isotherms (**a**) and streamlines (**b**) for ratio $R_i = R_o/2.6$ ($\Gamma = 3000$ g).

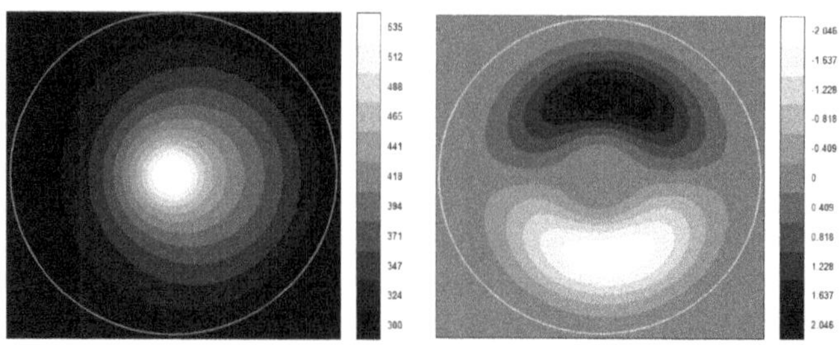

(a) (b)

Figure 7. Isotherms (**a**) and streamlines (**b**) for ratio $R_i = R_o/10$ ($\Gamma = 1000$ g).

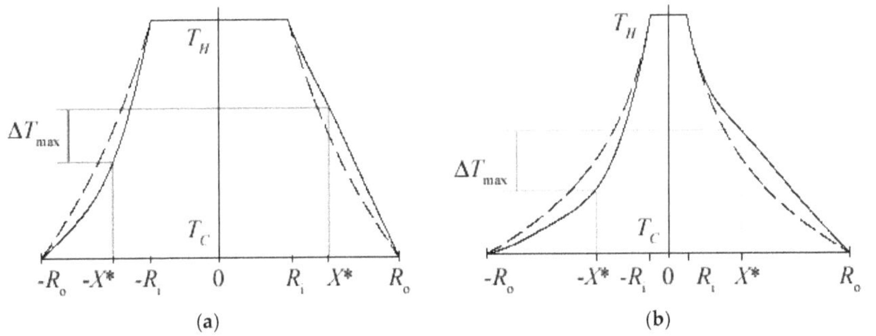

Figure 8. Temperature cross sections along OX axis for different ratios: $R_i = R_o/2.6$ and $\Gamma = 3000$ g (solid line), $\Gamma = 0$ (dashed line) (**a**); $R_i = R_o/10$ and $\Gamma = 1000$ g (solid line), $\Gamma = 0$ (dashed line) (**b**).

Our results were compared with a well-known asymptotic solution obtained by Hodnett [24] and Mack and Bishop [25]:

$$T(r,\varphi) = T(r) + (T_H - T_C)\mathrm{Ra}f\left(\frac{r}{R_i}\right)\sin\varphi. \quad (41)$$

Here the first, "diffusive", term is defined by Equation (40) and the second, "convective", term is expressed by a rather cumbersome formula for function f with a set of coefficients presented, for example, in [24].

It observed a good correspondence between the results obtained by means of our semi-analytical approach on the base of MQ-RFM and representation (41) (within the restrictions imposed on the asymptotic solution: the inner cylinder radius R_i and the temperature difference $(T_H - T_C)/T_C$ should be small enough). Our method demonstrated a very high rate of convergence. For example, in the case $R_i = R_o/2.6$, the relative error between approximate and analytical solutions became less than $1 \cdot 10^{-2}$ after only two iterations. Here, $\overline{\mathrm{Nu}}\big|_{\partial\Omega_C} \approx 242.5$, $\overline{\mathrm{Nu}}\big|_{\partial\Omega_H} \approx 238.6$, with relative errors between $1 \cdot 10^{-2}$ and $3 \cdot 10^{-2}$ respectively as compared with those evaluated with using expression (41) which gives us $\overline{\mathrm{Nu}}\big|_{\partial\Omega_C} = \overline{\mathrm{Nu}}\big|_{\partial\Omega_H} = 245.9$.

The obtained temperature profiles are also in good accordance with both experimental and numerical results obtained in the work of Garraud [36]. However, in her work, to obtain an appropriate accuracy, either a uniform finite-element grid must be used with 11,000 nodes or an adaptive non-uniform grid with 2000 nodes. At the same time, our approach provides a semi-analytical solution with the same accuracy in the form of a series of 36 terms only with respect to very simple 2D MQ-RBFs (38).

Beside temperature profiles, the temperature difference between two opposite points along the sensitivity axis OX was evaluated (Figure 9) to find an optimum relative position $X* = X(\Delta T_{\max})$ of sensors of a TA with respect to the inner radius. The relative position is as follows:

$$\frac{X^* - R_i}{R_o - R_i} \approx 0.375.$$

The sensitivity S was evaluated as a function of temperature difference $\Delta T_{\max}$ from external acceleration $\Gamma = \Gamma(\mathrm{g})$, where g is acceleration of gravity ($g \approx 9.8$ m/s^2) (Figure 10). A linear part of these dependence between $\Gamma = 0$ and $\Gamma = 1000$ g corresponds to the range of "small" and "medium" accelerations and shortens as R_i decreases.

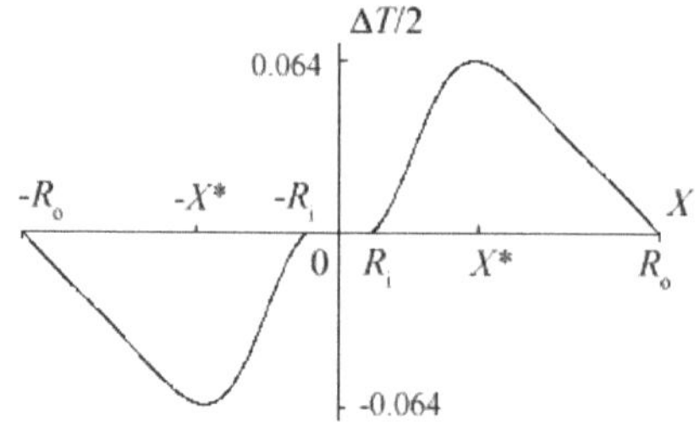

Figure 9. Temperature difference between two opposite points along the sensitivity axis OX ($R_i = R_o/10$); optimum position corresponds to X^* and $-X^*$.

It does not depend on Ra number and almost coincides with the theoretical estimation according to Hodnett's model (41) as well as to numerical and experimental data [36].

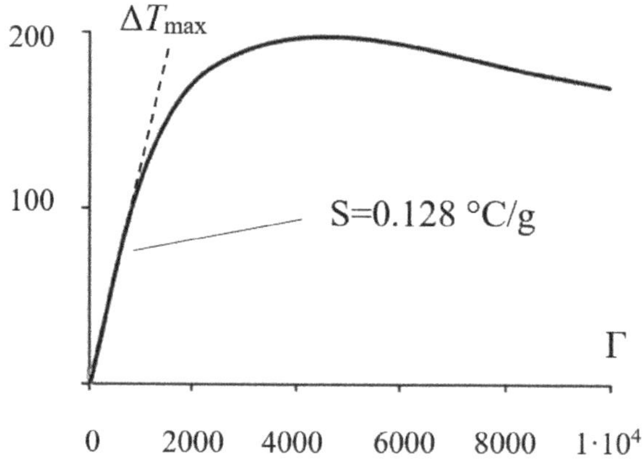

Figure 10. Plot of $\Delta T_{\max}$ vs. external acceleration $\Gamma = \Gamma(g)$ ($R_i = R_o/10$); the sensitivity S on the linear part between 0 and 1000 is about 0.128 °C/(m/s^2).

The same shape of the plots, with the steep linear slope in the range of "small" accelerations, maximum achieved at "medium" accelerations, and smooth deceleration at "high" accelerations, was observed experimentally [36]. In that work, the sensitivity S = 0.157 °C/g was measured in the interval between $\Gamma = 0$ and $\Gamma = 500$ g for the sensor with $R_o = 750$ µm, $R_i = 50$ µm. It should be noted that increase of acceleration to the values greater than some hundred g yields the turbulence phenomenon [46] which can explain partially the behavior of $S(\Gamma)$ plot. However, the simplified RFM-Galerkin scheme proposed in this article has low accuracy for such case and requires further modification to simulate turbulent flows. The results shown in Figure 10 and corresponding to values $\Gamma > 1000$ g can only be considered qualitatively.

7. Conclusions

The new modification of the Galerkin method for solving stationary convection-diffusion problems in arbitrarily-shaped domains was proposed. It is based on the combined use of the RFM with Boolean representation of the domain boundary and the Petrov-Galerkin iteration procedure with multiquadric RBFs. Numerical experiments showed the high accuracy and high rate of convergence of the novel approach. The semi-analytical solution was obtained in a closed form of the series with respect to MQ-RBFs and it satisfied the boundary conditions exactly. The technique was applied to the well-studied benchmark problem of convection in the circular annulus, which is the simplest model of the thermal accelerometer. The obtained results were in good accordance with experimental data, numerical and asymptotic solutions. MQ-RFM can be applied directly for evaluation of thermal fields in more complicated domains and easily generalized to the case of other types of boundary conditions including mixed ones in different parts of the boundary. For analytical investigation of bandwidth, a transient study based on a combination of MQ-RFM and the Rothe method can be realized by analogy with [11]. As for solving 3D problems, it should be noted that, with the help of RFM it is possible to describe any 3D object geometry. However, we cannot directly transfer the technique described in the article to 3D problems due to difficulties in extending the vorticity-stream function formulation to the multidimensional case [47].

Author Contributions: Conceptualization, A.G.; methodology, M.B.; validation, P.C.; formal analysis, M.B.; investigation, M.B.; resources, A.G.; data curation, P.C.; writing—original draft preparation, M.B.; writing—review and editing, A.G.; visualization, M.B.; supervision, A.G.; project administration, A.G. All authors have read and agreed to the published version of the manuscript.

Funding: This research received no external funding.

Conflicts of Interest: The authors declare no conflict of interest.

References

1. Garraud, A.; Giani, A.; Combette, P.; Charlot, B.; Richard, M. A dual axis CMOS micromachined convective thermal accelerometer. *Sens. Actuators A Phys.* **2011**, *170*, 44–50. [CrossRef]
2. Roache, P. *Computational Fluid Dynamics*; Hermosa Publishers: Albuquerque, NM, USA, 1972.
3. Patankar, S. *Computation of Conduction and Duct Flow Heat Transfer*; Innovative Research: Maple Grove, MN, USA, 1991.
4. Shih, T.-M. *Numerical Heat Transfer*; Hemisphere Publishing Corporation: Washington, DC, USA, 1984.
5. Wrobel, L.C.; Aliabadi, M.H. *The Boundary Element Method*; John Wiley & Sons: New York, NY, USA, 2002.
6. Fletcher, C. *Computational Galerkin Methods*; Springer Verlag: New York, NY, USA, 1984.
7. Katz, A.J. Meshless Methods for Computational Fluid Dynamics. Ph.D. Thesis, Department of Aeronautics and Astronautics, Stanford University, Stanford, CA, USA, 2009.
8. Kosec, G.; Šarler, B. Numerical Solution of Natural Convection Problems by a Meshless Method. In *Convection and* Conduction Heat Transfer; Ahsan, A., Ed.; InTech: Rijeka, Croatia, 2011.
9. Liu, G.R. *Mesh Free Methods: Moving Beyond the Finite Element Method*; CRC Press: Boca Raton, FL, USA, 2009.
10. Rvachev, V.L.; Sheiko, T.I.; Shapiro, V.; Tsukanov, I. On completeness of RFM solution structures. *Comput. Mech.* **2000**, *25*, 305–317. [CrossRef]
11. Matveev, V.A.; Basarab, M.A. Numerical modeling of heat diffusion processes in the solid-state wave gyro resonator by the R-function method. *Vestn. J. Bauman Mosc. State Tech. Univ. Nat. Sci. Eng.* **2005**, *1*, 98–110.
12. Artyukh, A. Mathematical and Numerical Modeling of Natural Convection in an Enclosure Region with Heat-conducting Walls by the R-functions and Galerkin Method. *Radioelectron. Inform.* **2012**, *4*, 103–108.
13. Suvorova, I.G.; Kravchenko, O.V.; Baranov, I.A. Mathematical and Computer Modeling of Axisymmetric Flows of an Incompressible Viscous Fluid by the Method of R-Functions. *J. Math. Sci.* **2012**, *184*, 165–180. [CrossRef]
14. Tsukanov, I.; Shapiro, V.; Zhang, S. A Meshfree Method for Incompressible Fluid Dynamics Problems. *Int. J. Numer. Methods Eng.* **2003**, *58*, 127–158. [CrossRef]
15. Basarab, M.A.; Matveev, V.A. A Meshless Semi-Analytical Method for Solving Convection Problems in Complex-Shaped Closed Cavities. *Appl. Mech. Mater.* **2015**, *763*, 170–174. [CrossRef]
16. Kansa, E.J. Multiquadrics–a scattered data approximation scheme with applications to computational fluid dynamics: I. Surface approximations and partial derivative estimates. *Comput. Math. Appl.* **1990**, *19*, 127–145. [CrossRef]
17. Kansa, E.J. Multiquadrics–a scattered data approximation scheme with applications to computational fluid dynamics: II. Solutions to parabolic, hyperbolic, and elliptic partial differential equations. *Comput. Math. Appl.* **1990**, *19*, 147–161. [CrossRef]

18. Ho-Minh, D.; Mai-Duy, N.; Tran-Cong, T. A Galerkin-RBF approach for the streamfunction-vorticity-temperature formulation of natural convection in 2D enclosured domains. *CMES Comput. Modeling Eng. Sci.* **2009**, *44*, 219–248. [CrossRef]
19. Ding, H.; Shu, C.; Yeo, K.S. Simulation of natural convection in eccentric annuli between a square outer cylinder and a circular inner cylinder using local MQ-DQ method. *Numer. Heat Transf. Part A* **2005**, *47*, 291–313. [CrossRef]
20. Basarab, M.A. Application of Voronoi diagrams for solving boundary value problems by means of the R-functions and RBF neural networks. In *Voronoi's Impact on Modern Science*; Book 4; Laurincikas, A., Steuding, J., Eds.; Institute of Mathematics, NAS of Ukraine: Kiev, Ukrayina, 2008; Volume 1, p. 260.
21. Atayilmaz, Ş.Ö. Experimental and numerical study of natural convection heat transfer from horizontal concentric cylinders. *Int. J. Therm. Sci.* **2011**, *50*, 1472–1483. [CrossRef]
22. Bishop, E.H.; Carley, C.T. Photographic studies of natural convection between concentric cylinders. *Proc. Heat Transf. Fluid Mech. Inst.* **1966**, 63–78.
23. Kuehn, T.H.; Goldstein, R.J. An experimental and theoretical study of natural convection in the annulus between horizontal concentric cylinders. *J. Fluids Mech.* **1976**, *74*, 695–719. [CrossRef]
24. Hodnett, P.F. Natural convection between horizontal heated concentric circular cylinders. *Z. Angew. Math. Phys.* **1973**, *24*, 507–516. [CrossRef]
25. Mack, L.R.; Bishop, E.H. Natural convection between horizontal concentric cylinders for low Rayleigh numbers. *Quart. J. Mech. Appl. Math.* **1968**, *21*, 223–241. [CrossRef]
26. Abbott, M.R. A numerical method for solving the equations of natural convection in a narrow concentric cylindrical annulus with a horizontal axis. *Quart. J. Mech. Appl. Math.* **1964**, *17*, 471–481. [CrossRef]
27. Charrier-Mojtabi, M.C.; Mojtabi, A.; Caltagirone, J.P. Numerical solution of a flow due to natural convection in horizontal cylindrical annulus. *J. Heat Transf.* **1979**, *101*, 171–173. [CrossRef]
28. Cho, C.H.I.; Chang, K.S.; Park, K.H. Numerical Simulation of Natural Convection in Concentric and Eccentric Horizontal Cylindrical Annuli. *J. Heat Transf.* **1982**, *104*, 624–630. [CrossRef]
29. Hassan, A.K.; Al-Lateef, J.M.A. Numerical Simulation of two dimensional transient natural convection heat transfer from isothermal horizontal cylindrical annuli. *J. Eng.* **2007**, *13*, 1429–1444.
30. Mizushima, J.; Hayashi, S.; Adachi, T. Transitions of natural convection in a horizontal annulus. *Int. J. Heat Mass Transf.* **2001**, *44*, 1249–1257. [CrossRef]
31. Powe, R.E.; Carley, C.T.; Bishop, E.H. Free convection flow patterns in cylindrical annuli. *J. Heat Transf.* **1969**, *91*, 310–314. [CrossRef]
32. Shu, C. Application of differential quadrature method to simulate natural convection in a concentric annulus. *Int. J. Numer. Methods Fluids* **1999**, *30*, 977–993. [CrossRef]
33. Yunlong, H. The Numerical Study of The Flow and Heat Transfer between Two Horizontal Cylinders. Master's Thesis, National University of Singapore, Singapore, 2004.
34. Lin, L. Design and Analysis of Microthermal Accelerometer. Ph.D. Thesis, Simon Fraser University, Burnaby, BC, Canada, 2004.
35. Bo Wu, A. Thermal Drift in Convective Micromachined Accelerometers. Master's Thesis, Simon Fraser University, Burnaby, BC, Canada, 2004.
36. Garraud, A. A Melioration des Performances et Nouveaux Concepts de Detecteurs de Capteurs Inertiels a Detection Thermique. Ph.D. Thesis, Universite Montpellier, Montpellier, France, 2011.
37. Giani, A.; Combette, P.; Deblonde, A.; Garraud, A.; Matveev, V.A.; Basarab, M.A. A Highly Sensitive Thermal Accelerometer with Improved Bandwidth. In Proceedings of the 20th Saint-Petersburg International Conference on Integrated Navigation Systems, Saint-Petersburg, Russia, 27–29 May 2013; pp. 191–198.
38. Kantorovich, L.V.; Krylov, V.I. *Approximate Methods of Higher Analysis*; Interscience Publishers: New York, NY, USA, 1958.
39. Rektorys, K. *Variational Methods in Mathematics, Science and Engineering*; Dr. Reidel Publishing Co.: Boston, MA, USA, 1980.
40. Rvachev, V.L.; Sheiko, T.I.; Shapiro, V.; Tsukanov, I. Transfinite interpolation over implicitly defined sets. *Comput. Aided Geom. Des.* **2001**, *18*, 195–220. [CrossRef]
41. Sidorov, M.V. Construction of structures for solving the Stokes problem. *Radioelectron. Inform.* **2002**, *3*, 52–54. (In Russian)

42. Sidorov, M.V. Application of R-functions to the calculation of the Stokes flow in a square cavity at low Reynolds. *Radioelectron. Inform.* **2002**, *4*, 77–78. (In Russian)
43. Michlin, S.G. *Variational Methods in Mathematical Physics*; Pergamon Press: Oxford, UK, 1964.
44. Chang, K.S.; Won, Y.H.; Cho, C.H. Patterns of Natural Convection around a Square Cylinder Placed Concentrically in a Horizontal Circular Cylinder. *J. Heat Transf.* **1983**, *105*, 273–280. [CrossRef]
45. Jehhef, K.A.; Badawy, F.A. Effect of inner boundaries geometry on natural convection heat transfer in horizontal annuli. *J. Eng.* **2010**, *16*, 6048–6063.
46. Chung, T.J. *Computational Fluid Dynamics*; Cambridge University Press: Cambridge, UK, 2006.
47. Ern, A.; Guermond, J.-L.; Quartapelle, L. Vorticity-velocity formulations of the Stokes problem in 3D. *Math. Methods Appl. Sci.* **1999**, *22*, 531–546. [CrossRef]

applied sciences

MDPI

Article

Development of Driftwood Capture Trellis for Capturing Driftwood in Agricultural Drainage Ditches

Youngseok Song [1] and Moojong Park [2,*]

1 Department of Civil Engineering and Landscape Architectural, Daegu Technical University, Daegu 42734, Korea; kind711@hanmail.net

2 Department of Aeronautics and Civil Engineering, Hanseo University, Seosan 31962, Korea

* Correspondence: mjpark@hanseo.ac.kr; Tel.: +82-416601051

Received: 7 July 2020; Accepted: 18 August 2020; Published: 21 August 2020

check for updates

Abstract: The flow of driftwood and soil into drainage from agricultural areas accelerates sedimentation and inflicts overflow damage after rainfall events due to insufficient discharge capacity, causing flooding on agricultural land. However, there have been few efforts to develop a driftwood capture trellis for agricultural drainage ditches, except for some suggested design criteria. In this study, we developed a driftwood capture trellis to capture driftwood in agricultural drainage ditches and evaluated its performance based on hydraulic characteristics. The facility was designed considering criteria for drainage and driftwood control barriers, as well as the properties of driftwood found near agricultural drainage ditches. Performance evaluation was conducted through hydraulic experiments. Driftwood capture trellises were installed in 400 mm drainage pipes and a total of 216 experimental runs were conducted: six runs each in six different velocity variations and six water depth variations. The results showed that the driftwood capture efficiency of the facility exceeded 60% at a velocity of 0.144 m^3/s. Limited conditions for hydraulic experiments should be considered. The driftwood capture trellis for agricultural drainage ditches developed in this study could contribute to a reduction in overflow damage caused by driftwood sedimentation.

Keywords: agricultural drainage ditches; driftwood; driftwood capture trellis; capture efficiency; hydraulic experiment; hydrodynamics

---

1. Introduction

Sedimentation of soil and debris into drainage pipes in urban and mountainous areas caused by local and sporadic rainfall may cause overflow damage if the discharge capacity of the drainage pipes is insufficient. The increased soil moisture resulting from drainage overflow increases the possibility of a disaster related to the flow of debris, thus increasing the risk of damage to life and property. A driftwood capture trellis or design standard for drainage facilities that protects against such damage should be considered; however, reduction measures related to drainage have not yet been established, domestically or internationally. Drainage systems are used in agricultural, urban, and mountainous areas that vary in size depending on intended use. Sudden local downpours of rain tend to occur more frequently in recent years due to climate change, and cause continuous damage despite various disaster prevention and reduction measures [1–6]. Design criteria for drainage systems have been established both domestically and internationally, but little research has been conducted on the development of driftwood capture trellises, as these facilities are mostly small-scale [7–10]. From 2009 to 2018 in the Republic of Korea, climate change increased rainfall, flooding 22,323 ha of farmland and inflicting damage on a 0.61 ha to 13,154 ha area. The main culprit of the farmland flooding was overflow caused by a lack of discharge capacity because of the sedimentation of driftwood and soil in the drainage system [11]. Therefore, formulation of measures to develop and maintain effective driftwood capture trellises and increase the discharge capacity of agricultural drainage ditch systems is an urgent concern.

Drainage research has mostly been centered around hydraulic experiments or numerical modeling. Empirical equations have been suggested to calculate the amount of discharge flowing through drainage systems, which take into account the longitudinal slopes of roadways, transverse slopes, local constants, and changes in interception capacity [12–15]. As for road drainage, hydraulic experiments were conducted to analyze the interception efficiency and the effects of sediment concentration and bed slope on driftwood flow deposition and runoff reduction [16–20]. Some researchers calculated nonuniform flow in drainage systems through numerical modeling or assessed the safety of installing driftwood capture trellises [21,22]. Previous research on drainage mostly focused on the calculation of inflow considering discharge, sedimentation, and flow characteristics, but the efficacy of drainage driftwood capture trellises and the effect of a lack of discharge capacity have not been adequately addressed.

Previous studies on driftwood capture trellises have been concentrated on hydraulic experiments or design methods for large-scale facilities related to landslide hazards, rather than on smaller drainage systems. The research has covered hydraulic characteristics, including design methods for bridges or culverts that aim to reduce driftwood, analysis of capture efficiency based on the size of driftwood, and regression equations for design improvements [23–30]. Some researchers have explored the characteristics of driftwood sedimentation locations and capture efficiency based on structural changes and screen spacing in permeable driftwood control barriers [31–34]. The capture efficiency of debris fins, screens, and deflector screens, among other debris control tools suggested by the Federal Highway Administration (FHWA [35]), have been analyzed; all except debris fins were found to be highly effective in reducing debris flow [36].

An examination of the drainage design criteria of different nations revealed that they only provide installation criteria, not standards for the operation of facilities to mitigate disasters. In large-scale drainage systems which have high-frequency designations, driftwood capture trellis debris control facilities or retention ponds can be established, but this is not common for most smaller drainage systems located in agricultural and mountainous areas. The U.S. Department of Transportation Federal Highway Administration (FHWA [35]) and the Ministry of Land, Infrastructure, and Transport of Korea recommend the installation of debris control facilities to keep debris out of storm drains connected to the drainage system [35,37]. Debris control structures are structures placed across well-defined channels to form basins which impede drainage flow. Practices that reduce the quantity of floating debris include directional felling uphill with a tree-pulling system and providing a buffer strip of undisturbed vegetation along the drains. Debris control structures are divided into six groups according to type of debris, as shown in Figure 1.

Figure 1. Debris control structures: (**a**) deflector; (**b**) rack; (**c**) riser; (**d**) crib; (**e**) fin; (**f**) dam [35].

Debris can be grouped based on size into the following categories: light floating debris (LFD), medium floating debris (MFD), heavy floating debris (HFD), floating debris (FD1), fine detritus (FD2), coarse detritus (CD), and boulders (Table 1). Debris in the drainage from agricultural lands or plains is expected to include smaller twigs, dead leaves, grit, and sand; it is difficult to classify or define clear standards for the classification of such debris. According to the standards of the FHWA [35], LFD can be defined as debris found in drainage from plains and agricultural areas, while other types are found in mountainous areas and near streams [35]. The debris control structures suggested in the past were mostly for large-scale disaster mitigation facilities in mountainous areas and near streams. There is an urgent need for the development of debris driftwood capture trellises in small-scale drainage systems in plains and agricultural lands.

Table 1. Type of debris.

Debris Classification	Type
Light Floating Debris (LFD)	small twigs, wood chips, cloth
Medium Floating Debris (MFD)	twigs, big wood chips
Heavy Floating Debris (HFD)	log, timber
Floating Debris (FD1)	fluent substances, including clay, silt, sand, pebble, wood chips
Fine Detritus (FD2)	homogeneous silt, sand, and gravel without suspended solids deposited where flow is slow
Coarse Detritus (CD)	coarse gravel, rack fragments
Boulders (B)	boulder, large rock fragment in the event of a flood

Many researchers have studied large-scale facilities, including permeable driftwood control barriers or facilities, in an attempt to alleviate the damage produced by sediment-related disasters. However, driftwood reduction methods in large-scale facilities, which mostly deal with logs, not twigs, in drainage systems, are of limited applicability to smaller facilities. In general, too little attention has been paid to driftwood capture trellises for agricultural drainage ditches, both domestically and internationally.

This study intends to develop a driftwood capture trellis that captures driftwood from agricultural drainage ditches. Using a hydraulic experiment, the study estimates driftwood capture efficiency under a variety of experimental conditions and establishes installation standards for agricultural drainage ditches. In addition, it aims to determine whether design discharge is installed and estimate the capture efficiency in various areas, by developing formulas that estimate discharge conditions and capture efficiency.

2. Materials and Methods

2.1. Drainage Design Criteria

Drainage refers to ditches created to drain water. Drainage is applicable to various facilities and is installed according to the recommended frequency (not based on size). Drainage systems are used as complementary facilities to culverts, roads, streets, and slopes. Different drainage design criteria exist depending on location and expected frequency, from a 2-year frequency to a maximum 100-year frequency (Table 2). The United States divides drainage systems into natural, major, and minor drainage methods, and plans around the occurrence of 2- to 100-year events [38,39]. Japan plans around 20- to 50-year events, based on its geographical characteristics and climate conditions [40]. In Canada, drainage design plans for a 2- to 100-year event depending on the geographical characteristics of the terrain in which drainage systems are installed [41]. In Australia, the frequency varies from 3 to 40 years depending on the purpose of the specific drainage system [42]. Korea classifies drainage systems into mountainous areas, plains, and agricultural lands and sets the frequency as 20 to 50 years for mountainous areas, 10 to 30 years for plains, and 20 years for agricultural lands. Additionally, it has been suggested that at least 20% of these systems should be adjusted based on estimated flood likelihood and in consideration of the sedimentation of driftwood and soil in drainage [7–9].

Table 2. Frequency for drainage by country.

Nation	Target Areas	Frequency
U.S.	All	2 to 100 years
Canada	All	2 to 100 years
Australia	All	3 to 40 years
Japan	All	20 to 50 years
Korea	Mountain Area	20 to 50 years
	Plain	10 to 30 years
	Agricultural Land	20 years

Drainage systems can be installed in a wide range of locations, including mountainous areas, plains, and farmlands. The recommended frequency is higher for drainage systems in agricultural and mountainous areas compared to plains and roads. This is because the probability of driftwood and soil inflow into the drainage is higher in agricultural and mountainous areas than in plains and roads. However, no adequate guidelines have been provided for installing and managing driftwood capture trellises to alleviate the damage from insufficient discharge capacity due to this increased driftwood flow. Against this backdrop, it is necessary to create guidelines for development and management of driftwood capture trellises in agricultural drainage ditch systems.

2.2. *Definition of Driftwood in Agricultural Drainage Ditches*

Regarding debris in agricultural drainage ditches, there are no clear standards for size or type except the suggestions of the FHWA [35], which classifies small twigs and wood chips as LFD [35]. This study presents a definition of driftwood in agricultural drainage ditches in order to help develop a driftwood capture trellis. The study looks at agricultural drainage ditches in the plains region in Chungcheongnam-do, Korea. Topographically, Chungcheongnam-do is the lowest area in Korea, with an average altitude of less than 100 m. It has an average annual temperature of 11–13 °C, and average annual rainfall of 1100–1350 mm. The basin area of agricultural drainage ditches where driftwood is collected is 3.67 ha. Agricultural drainage ditches located in the target region are square-shaped with a width of 400 mm and are installed in concrete. The circumference of the target region is about 1 km, and agricultural drainage ditches are installed around 600 m. For this study, twigs and wood chips were collected within 1 km of the agricultural area, in and around the location of agricultural drainage ditches (Figure 2).

In the agricultural drainage ditches area, driftwood from various plants and crops were collected that fell within a size range that would be able to flow into the agricultural drainage ditches. A total of 350 pieces of driftwood were collected that were 1 to 25 mm in diameter and 100 to 500 mm in length. More than 60% of the pieces of driftwood were within 2 to 6 mm in diameter. In addition, 90% of the collected driftwood specimens were 12 mm or less in diameter, and only about 10% were between 12 mm and 25 mm in diameter. The driftwood entering agricultural drainage ditches was confirmed to be mostly light floating driftwood (LFD). The details of the driftwood collected are shown in Table 3.

Figure 2. Driftwood found near agricultural drainage ditches.

Table 3. Driftwood found near agricultural drainage ditches.

Diameter (mm)	Count	Ration (%)	Diameter (mm)	Count	Ration (%)
1–2	16	4.6	12–13	6	1.7
2–3	46	13.1	13–14	5	1.4
3–4	65	18.5	14–15	6	1.7
4–5	59	16.8	15–16	2	0.6
5–6	44	12.6	16–17	1	0.3
6–7	34	9.7	17–18	3	0.9
7–8	16	4.6	18–19	2	0.6
8–9	16	4.6	19–20	1	0.3
9–10	9	2.6	20–25	4	1.1
10–11	7	2.0	Sum	350	100.0
11–12	8	2.3			

2.3. Development of Driftwood Capture Trellis for Agricultural Drainage Ditches

Few researchers have paid attention to the development of driftwood flow of driftwood capture trellises for agricultural drainage ditch systems. Most previous research has focused on large-scale damage mitigation facilities, including driftwood control structures or ring net barriers; smaller agricultural drainage ditch systems have not received sufficient attention. In this study, we examine drainage design criteria, driftwood control structures, and the characteristics of driftwood in order to develop a driftwood capture trellis for agricultural drainage ditches.

The study incorporated diverse standards for the driftwood flow of driftwood capture trellises, as follows. First, the installation of a driftwood capture trellis in agricultural drainage ditches was assumed to cause overflow damage due to the deposition of driftwood. To prevent overflow, we set the overflow height at 20% of the drainage height, as suggested in established design criteria. Second, the standard for the driftwood to be reduced in agricultural drainage ditches was established at 12 mm or less in diameter. Third, the opening, which accounts for 80% of the drainage height in an agricultural drainage ditch system (the remainder is overflow), was designed with grates similar to those in driftwood flow of driftwood capture trellises. The horizontal and vertical net distance standards followed those of the design criteria for permeable driftwood barriers in Japan as there are no relevant existing criteria for general driftwood control facilities [43]. Permeable driftwood barriers are installed to capture driftwood flow and driftwood in sediment disasters, and the design criteria for the opening of the facility is set at 1.0 times the maximum diameter for the horizontal and vertical net distance. However, driftwood control barriers, which are large structures designed to capture driftwood or rocks, can be blocked by driftwood when used in smaller facilities such as agricultural drainage ditch systems. Thus, smaller facilities set the design criteria at 1.0 to 2.0 times the maximum diameter for the horizontal and vertical net distance. Since the driftwood collected around the agricultural drainage ditches had a maximum diameter of 25 mm, the horizontal and vertical net distance of the opening was set from 25 to 50 mm.

As mentioned above, the driftwood capture trellis for agricultural drainage ditches suggested in this study was developed considering various design criteria. The parameters of the agricultural drainage ditch system are as follows: drainage width (B), drainage height (H), opening height (h), overflow height (y), horizontal net distance (d1), and vertical net distance (d2). A schematic design of the driftwood capture trellis is shown in Figure 3.

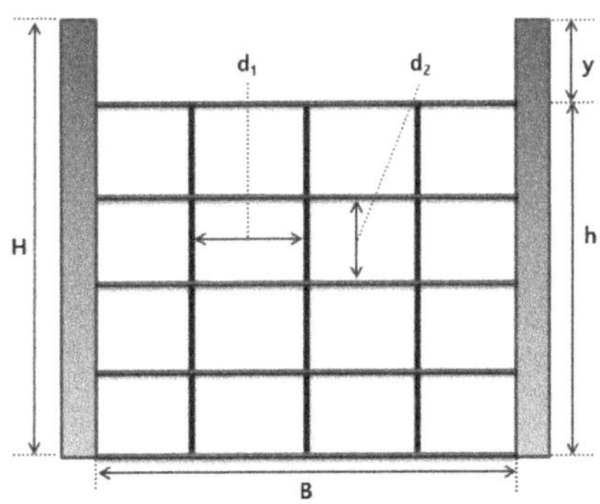

Figure 3. Design of the driftwood capture trellis for agricultural drainage ditches.

2.4. Manning Equation

The Manning equation is used to estimate the average velocity in open channels and pipe conduits. The roughness coefficient determined by the quality of riverbeds was applied by eliminating the depth effect of the Chezy coefficient. In addition, n is the roughness coefficient of the Manning equation, which is often determined by the quality and forms of riverbeds, but is rarely affected by flow characteristics. The Manning equation is shown in Equation (1).

$$V = \frac{1}{n} {R_h}^{2/3} {S_O}^{1/2} \quad (1)$$

where, V is average velocity, n is the roughness coefficient, R_h is hydraulic radius, and S_o is the slope of the hydraulic grade line. The depth estimated by the Manning equation is called the normal depth, and the riverbed slope at the normal depth is called the normal slope.

2.5. Froude Number

The Froude number refers to the inertial force ratio relative to the gravity of flow or the surface velocity ratio relative to the average velocity of flow. Flow states are divided into ordinary, critical, and supercritical flow, depending on the ratio of inertial force to gravity. The formula used to estimate the Froude number is shown in Equation (2).

$$F_r = \frac{V}{\sqrt{gh}} \tag{2}$$

where, V is the local flow velocity, g is the gravitational acceleration, and h is the average depth of the channel section. If $Fr < 1$, the flow state is ordinary, and flow is dominated by gravity rather than by inertial force; such a flow, with relatively high depth and low velocity, seems to be normal. If $Fr > 1$, the flow state is supercritical, and is dominated by inertial force rather than by gravity; such a flow, with a relatively low depth and high velocity, seems to be non-normal.

3. Results

3.1. Characteristics of the Agricultural Drainage Ditches Hydraulic Lab

In prior studies, the specifications used for and the foundations of experiments in hydraulic laboratories were generally referenced, but details about the experimental set-up were limited. Furthermore, most hydraulic labs adopted scaled models to examine the hydraulic characteristics of large-scale facilities, which can confirm qualitative influences but are limited in their ability to show quantitative influences. In this study, a lab was built to conduct hydraulic experiments on an agricultural drainage ditch system at a scale of 1:1 to the actual drainage, as shown in Figure 4, so as to most clearly reveal the hydraulic characteristics of the drainage.

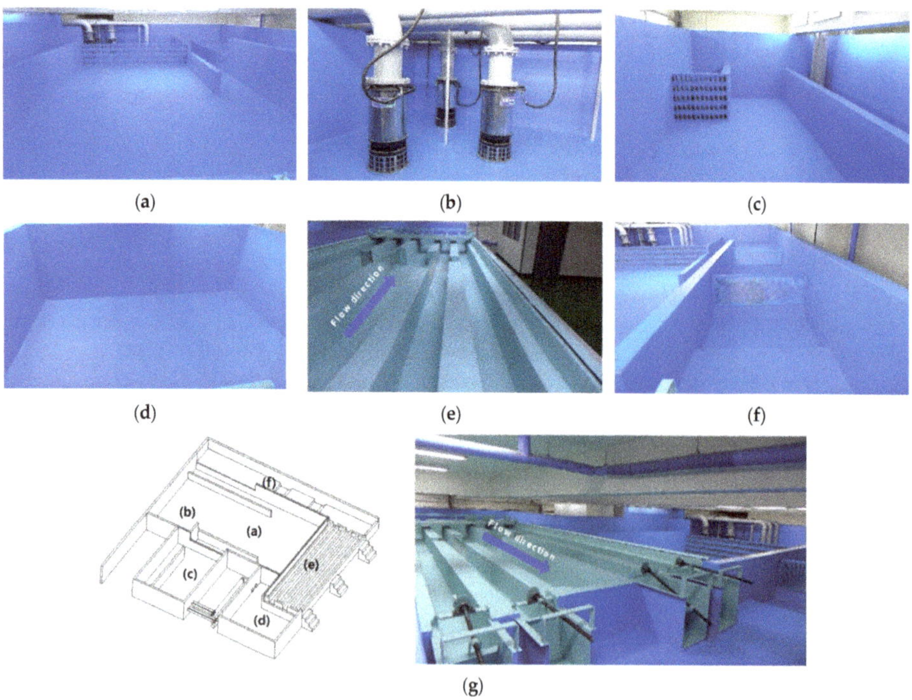

Figure 4. Hydraulic lab used to study agricultural drainage ditches: (**a**) water tank; (**b**) pump station; (**c**) head tank; (**d**) rectifying tank; (**e**) agricultural drainage ditches (400 mm); (**f**) return tank; (**g**) plan view.

The hydraulic laboratory has a total area of approximately 306.25 m^2 (17.5 m wide and long). Water from the water tank is transferred to the head tank through the pump, and water is supplied to the drainage through the rectifying tank to ensure a stable supply of water. The available discharge was used at a rate of 0.32 m^3/s with a width of 400 mm, a depth of 400 mm, and velocity of 2.0 m/s to model agricultural drainage ditches. The water used in the experiment was designed as a looping supply, since water is supplied to the water tank through the return tank. The entire system stores around 114 m^3 of water (in the water tank, head tank, and rectifying tank). The maximum supply of water was maintained for five minutes in the hydraulic experiment. The width of the agricultural drainage ditches was 400 mm, which was the average size of the drainage troughs in the surveyed agricultural areas. The amount of stored water was sufficient to supply the water flow for the hydraulic experiment. The specifications of the hydraulic experiment for each facility in the laboratory are shown in Table 4.

Table 4. Specifications for hydraulic laboratory.

No.	Facilities	Dimension (Width × Depth × Height)	Capacity
(a)	Water Tank	11,000 mm × 6850 mm × 700 mm	52.8 m^3
(b)	Pump Facilities	15 HP × 2ea, 10 HP × 1ea	0.33 m^3/s
(c)	Head Tank	4000 mm × 6000 mm × 2000 mm	48.0 m^3
(d)	Rectifying Tank	5000 mm × 4000 mm × 700 mm	14.0 m^3
(e)	Drainage	4000 mm × 8750 mm × 400 mm	1.4 m^3
(f)	Return Tank	14,000 mm × 1500 mm × 200 mm	4.2 m^3

3.2. Experimental Conditions

The conditions of the hydraulic experiment for agricultural drainage ditches were set for driftwood capture trellis, driftwood, and experimental conditions. Driftwood capture trellises for agricultural drainage ditches are areas that have not yet been studied at the domestic and international level. It is difficult to apply the same design criteria for large-scale driftwood capture trellises because design criteria for driftwood capture trellises for agricultural drainage ditches have not been established. Therefore, it is necessary to review additional experiments after preliminary application to existing unestablished research fields.

As mentioned above, the driftwood flow of the driftwood capture trellis designed in this study was developed considering various design criteria. A 400 mm wide agricultural drainage ditch was installed in the hydraulic laboratory along with a driftwood capture trellis with overflow height, opening height, and horizontal and vertical net distance as suggested in the design plan. The driftwood capture trellis was a square 400 mm in length, 80 mm in overflow height, 320 mm in opening height, 46 mm in horizontal net distance, and 35 mm in vertical net distance, with barriers 5 mm in diameter. The barriers were designed in an 8 × 8 grid, with horizontal and vertical net distances in the range of 1.0 to 2.0 times the maximum diameter of the driftwood, or between 25 and 50 mm. The driftwood capture trellis for agricultural drainage ditches developed in this study is shown in Figure 5.

It is difficult to apply driftwood collected in the vicinity of agricultural drainage ditches to various hydraulic experiments. In order to carry out many hydraulic experiments, it is necessary to apply the same sized driftwood in each. However, realistically, it is difficult to collect the same sized driftwood. Therefore, it is necessary to select an experimental tool that can provide the same size of driftwood and set it up according to the experimental conditions. Therefore, in this study, circular construction materials with various diameters were used. The opening of the agricultural drainage ditch pipe, excluding the overflow height, was 400 mm wide, 320 mm high, and had an area of 128,000 mm^2. In the experiment, 50% of the total opening area, or 64,000 mm^2, was applied as the size of one load of driftwood to test various velocity and water depth conditions. The length of the driftwood was set at 160 mm, or 50% of the opening height, and the driftwood had the shape of a round bar, similar to the twigs that were collected. The diameter of the driftwood was set at 12 mm or less, which represented

90% of the collected twigs, and there were 10 to 12 pieces of driftwood in one load with diameters of 3, 5, 7, 10, and 12 mm. The specifications of the driftwood used in the hydraulic experiments are shown in Table 5.

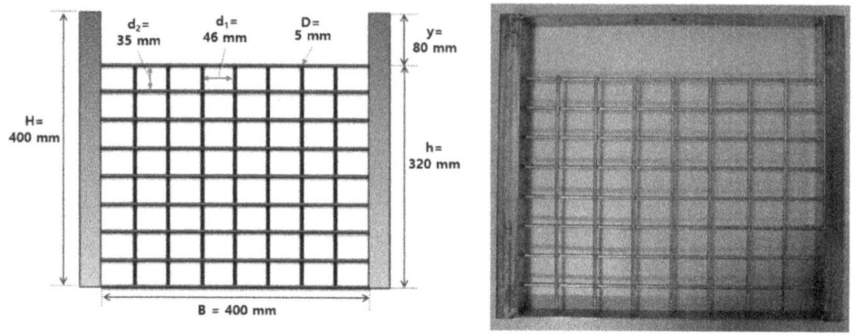

(a) (b)

Figure 5. Driftwood capture trellis for agricultural drainage ditches: (**a**) driftwood capture trellis (plan); (**b**) photograph of the implemented driftwood capture trellis.

Table 5. Driftwood types and specifications.

Driftwood	Length (mm)	Diameter (mm)	Count	Effective Cross-Sectional Area (mm^2)
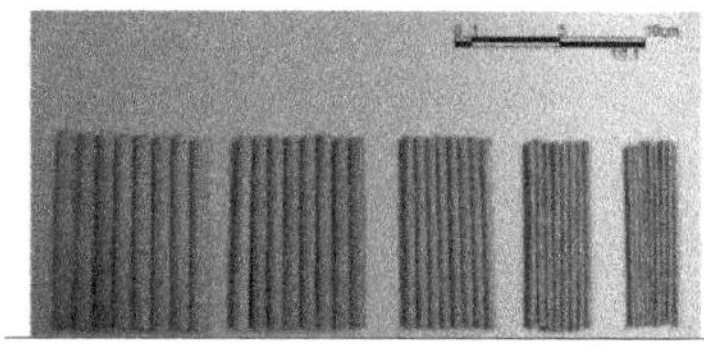	160	3	12	5760
		5	12	9600
		7	12	13,440
		10	10	16,000
		12	10	19,200
Sum			41	64,000

In the hydraulic experiments, the range of the discharge that can be supplied by the pump station was determined by the depth and velocity of the flow. The maximum discharge of the pump of the hydraulic laboratory was 0.33 m^3/s, assuming a pump efficiency of 80%, so the experimental conditions were set at 0.26 m^3/s or less, depending on the depth and velocity of the flow. As a result, six variations of the velocity (0.3, 0.6, 0.9, 1.2, 1.5, and 1.8 m/s) and six variations of the depth (0.08, 0.12, 0.16, 0.20, 0.24, and 0.28 m) were used. The discharge range was set at 0.01 to 0.20 m^3/s to meet the pump capacity based on the efficiency of the hydraulic laboratory. A total of 216 runs of the experiment were conducted in this study. The experimental conditions were as follows: one variation for agricultural drainage ditches, one variation for driftwood capture trellis, six variations for velocity, six variations for water depth, and six repetitions of driftwood drops, as shown in Table 6.

Table 6. Experimental conditions for agricultural drainage ditches.

Agricultural Drainage Ditches (Width(m))	Driftwood Capture Trellis	Velocity (m/s)	Depth (m)	No. of Drops	No. of Driftwood Drops
0.4	8 × 8 grid	0.3 0.6 0.9 1.2 1.5 1.8	0.08 0.12 0.16 0.2 0.24 0.28	6 drops	1 drop of driftwood = 41 pieces, ø3 mm ~ ø 12 mm in diameter
Number of experiment runs				216	

A digital point gauge (PH-355) was used to observe depth and a one-dimensional electronic hydrometer (FLO-MATE 2000) was used to observe velocity under the hydraulic experiment conditions (Figure 6). The depth was measured by observing the distance at which the needle at the lowest part of the point gauge reached the surface of the water. The total length of the observer was 600 mm and the valid measurement length was 400 mm, with an accuracy of ± 0.01 mm and a margin of error of ± 0.04 mm. Velocity was measured by observing the potential difference between two sections of fluid flow. The measurable velocity ranged from 0 to 2 m/s, and 1 directional characteristic of velocity was observed.

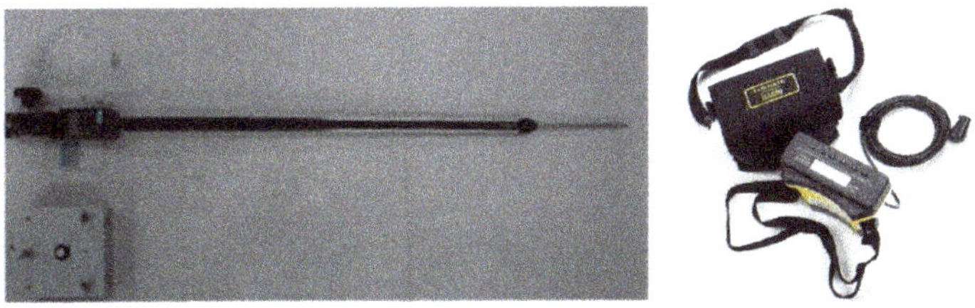

(**a**) (**b**)

Figure 6. Observation station: (**a**) digital point gauge; (**b**) one-dimensional electronic hydrometer.

3.3. Driftwood Capture Efficiency of the Driftwood Capture Trellis

Out of the 216 runs of the experiment, six runs were conducted for each of six variations of velocity and six variations of water depth—36 different variations in total. During the experiment, overflow damage was observed in three runs: when the water depth was 240 mm and the velocity was 1.8 m/s, and when the water depth was 280 mm and the velocity was 1.5 m/s and 1.8 m/s. Overflow damage was attributable to insufficient discharge capacity. The results from the experiments on minimal and maximal velocity at which driftwood is captured according to depth condition are shown in Appendix A. The driftwood capture efficiency of the driftwood capture trellis was between 50 and 100%, as shown in Figure 7: 80–100% at 80 mm water depth; 60–100% at 120 mm water depth; 50–100% at 160 mm water depth; 50–95% at 200 mm water depth; 70–95% at 240 mm water depth; and 80–97% at 280 mm water depth.

The average capture efficiency observed during the six runs of the hydraulic experiment for each of the 33 conditions is shown in Table 7. The observed capture efficiency, which ranged from 50.88 to 100%, tended to decrease as the velocity increased at the same water depth. Capture efficiency increased with increasing velocity, and there was a difference of less than 20% to more than 50% depending on the water depth condition. This is most likely due to fluctuations in capture efficiency according to flow and location of the driftwood capture trellis grid. The flow effect changed to a supercritical flow at velocities greater than 1.2 m/s, and the capture efficiency rapidly decreased. In order to consider the flow effect on both depth and velocity, the installation conditions of the driftwood capture trellis are presented in terms of discharge and Froude number.

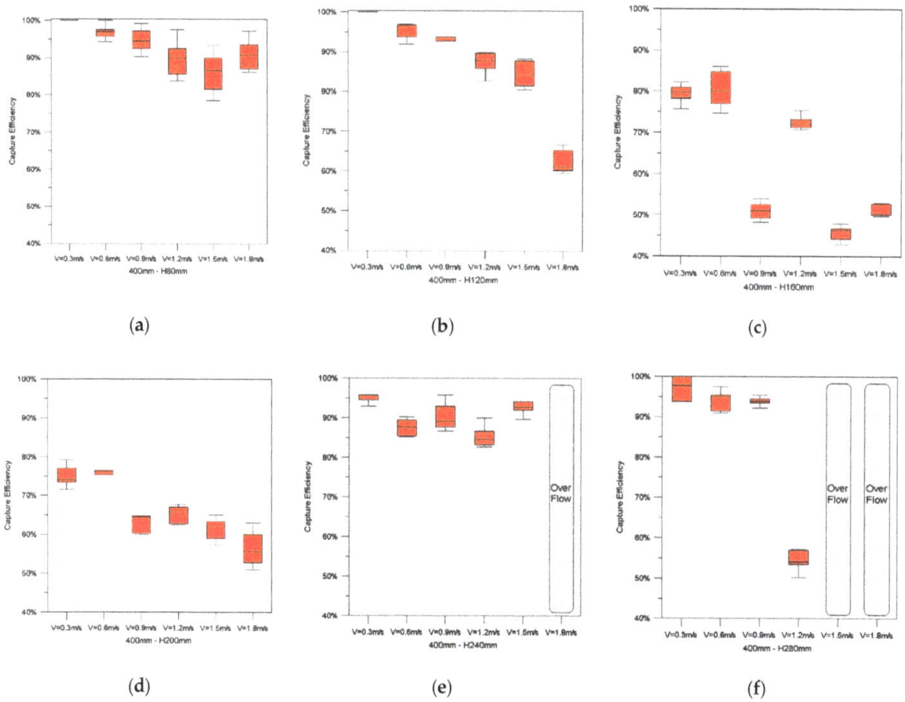

Figure 7. Capture efficiency of the driftwood capture trellis according to depth: (**a**) H = 80 mm; (**b**) H = 120 mm; (**c**) H = 160 mm; (**d**) H = 200 mm; (**e**) H = 240 mm; (**f**) H = 280 mm.

Table 7. Capture efficiency of the driftwood capture trellis by experimental conditions.

Water Depth (mm) / Velocity (m/s)	Unit (%)					
	80	**120**	**160**	**200**	**240**	**280**
0.3	100.00	100.00	99.44	94.69	95.44	97.31
0.6	96.81	95.94	90.44	86.13	87.63	94.31
0.9	94.69	93.13	81.00	86.50	89.81	93.94
1.2	89.50	87.75	82.19	75.31	84.81	81.70
1.5	86.13	84.38	75.88	71.69	67.94	overflow
1.8	80.50	62.00	50.88	52.25	overflow	overflow

3.4. Hydraulic Performance Evaluation of the Driftwood Capture Trellis

Design standards for hydraulic structures are created depending on the design discharge and the characteristics of the region in Korea. In addition, the size of an agricultural drainage ditch is set so that a stable flow occurs at the design discharge. In order to consider various discharge and flow conditions, an equation for calculating discharge and Froude number was proposed. Given the capture efficiency obtained for the velocity and depth of the flow, discharge capture efficiency curves and Froude number were estimated using the continuity and Manning equations (Figure 8).

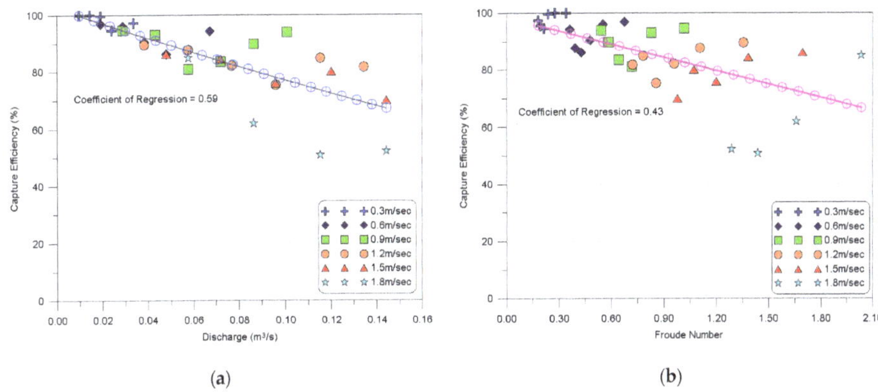

Figure 8. Capture efficiency of the driftwood capture trellis by hydraulic characteristics: (**a**) capture efficiency by discharge; (**b**) capture efficiency by Froude number.

The size of agricultural drainage ditch systems is planned considering various discharge rates. Therefore, in order to install a driftwood capture trellis for agricultural drainage ditches, a regression equation was calculated using the estimated curve to consider the effect of capture efficiency according to flow rate and flow state. The regression equation used to estimate the capture efficiency of the driftwood capture trellis was $CE = 101.6 - 243.8 \times Q$ for flow and $CE = 98.4 - 15.6 \times Fr$ for Fr. Here, CE is capture efficiency, Q is discharge, and Fr is Froude Number. According to the estimation curve, within the discharge range of 0.01 to 0.144 m^3/s, the capture efficiency of the driftwood capture trellis was ≈ 60% or greater. The less the rate of discharge, the higher the capture efficiency: 80% capture efficiency was maintained at 0.10 m^3/s or lower. As for the Froude number, the capture efficiency was estimated at ~80% or higher when $Fr < 1$, which represents a subcritical flow, and ≈ 60% or higher if $1 < Fr < 2.1$, which represents a supercritical flow.

The driftwood capture trellis was found to capture approximately 60% of the driftwood in agricultural drainage ditches according to discharge rate and Froude number. However, when the flow rate in the drainage was 0.144 m^3/s or greater, overflow occurred due to a reduction in the opening area caused by the captured driftwood. Based on these findings, the driftwood capture trellis in the agricultural drainage ditch system developed in this study would guarantee over 60% capture efficiency at a flow rate of 0.144 m^3/s or less. Moreover, using such hydraulic characteristics as flow rate and Froude number, it is possible to establish installation standards for driftwood capture trellis according to the target capture efficiency. The capture efficiency of the regression equation proposed in this study is based on data obtained in various experiments. Although qualitative results were calculated using the regression equation, as discharge and increase, the volatility of the capture efficiency greatly increases. It is necessary to install driftwood capture trellises in consideration of this uncertainty.

4. Discussion

Most existing driftwood capture trellises for capturing driftwood are large-scale, such as erosion control facilities in mountainous areas or deflectors that prevent driftwood from entering sewer systems; small facilities such as those designed for agricultural drainage ditches have not yet been developed. In Korea, damage has continuously occurred from 2009 to 2018 due to a lack of discharge capacity in agricultural drainage ditches [11]. In spite of this continuous damage, effective smaller driftwood capture trellises have not been developed, in part due to the insufficient effects of the small-scale facilities that do exist and, therefore, incessant maintenance issues.

Most previous studies studied the effects of driftwood capture trellises by using hydraulic experiments, according to size and form [12,13,16–20]. This study developed a driftwood capture trellis for agricultural drainage ditches and suggested the normal discharge rate as less than 0.144 m^3/s. The results of the hydraulic experiment showed that overflow damage occurred along with the capture of driftwood if the discharge rate exceeded 0.144 m^3/s after the installation of the driftwood capture trellis. A formula for estimating the driftwood capture efficiency according to discharge conditions was proposed for the hydraulic experiment, and a capture efficiency of 60–100% was estimated at a discharge rate of less than 0.144 m^3/s. The estimated capture efficiency in this study is higher than those seen in other studies based on hydraulic experiments [31–34].

It should be acknowledged that the facilities developed in this study have certain limitations. For example, they target driftwood flow without considering the influence of soil sedimentation. However, when driftwood and soil are combined, it is difficult to conduct hydraulic experiments or numerical modeling for driftwood capture trellises. Further studies are required to investigate the applicability of this type of agricultural drainage ditch facility and to assess the measures that must be taken to manage and maintain such facilities. In addition, the safety of the impact force of the driftwood capture trellis due to the inflow of driftwood must be considered. It is necessary to monitor actual rainfall events by installing the developed driftwood capture trellis at a test site. This will allow for establishment of maintenance standards considering the capture effect and impact force of driftwood from rainfall events.

It should be possible to maintain the effectiveness of driftwood capture trellises in continuous use. In addition, since the current study was limited to driftwood reduction, the influence of soil deposits in agricultural drainage ditches should be analyzed with numerical modeling in order to incorporate it in the development of future driftwood capture trellises for agricultural drainage ditches.

5. Conclusions

In this study, we developed a driftwood capture trellis to prevent overflow damage caused by the sedimentation of driftwood flowing into agricultural drainage ditches and evaluated the performance of the system through hydraulic experiments. According to the hydraulic experiment, a capture efficiency of 50.88–100.00% for driftwood in ditches in laboratory conditions was observed. Under several experimental conditions (i.e., discharge greater than 0.144 m^3/s), overflow damage was observed. The driftwood capture trellis developed in this study is thus expected to be installed at sites with rates of discharge less than 0.144 m^3/s.

The discharge was calculated according to the experimental conditions of water depth and velocity, and a regression equation for estimating the discharge condition and capture efficiency was proposed. The regression equation used to calculate the capture efficiency of the driftwood capture trellis considering discharge is $CE = 101.6 - 243.8 \times Q$. At a flow rate of 0.01 to 0.144 m^3/s, the capture efficiency was higher than 60% on the estimation curve, and the lower the flow rate, the higher the capture efficiency. The capture efficiency regression equation of the driftwood capture trellis considering Fr is $CE = 98.4 - 15.6 \times Fr$. The capture efficiency was estimated at about 80% or higher if $Fr < 1$, which represents a subcritical flow, and about 60% or higher if $1 < Fr < 2.1$, which represents a supercritical flow.

Based on the results of this study, the driftwood capture trellis will be useful in mitigating overflow damage by capturing driftwood in agricultural drainage ditches before it is deposited or flows into the drainage. Of course, there were limitations in terms of the number of experimental conditions used to develop the driftwood capture trellis, application of driftwood, and topographical characteristics. In addition, proper maintenance is required after installation of the driftwood capture trellis. Future research would greatly benefit from testing varying grid sizes to suit various topographical conditions and design facilities to meet a specific target capture efficiency.

Author Contributions: Conceptualization, Y.S. and M.P.; methodology, Y.S. and M.P.; validation, Y.S.; formal analysis: Y.S.; resources, Y.S. and M.P.; data curation, Y.S. and M.P.; writing-original draft preparation, Y.S.; writing-review and editing, M.P.; visualization, Y.S.; supervision, Y.S. and M.P.; project administration, M.P. All authors have read and agreed to the published version of the manuscript.

Funding: This work was funded by the Korea Meteorological Administration Research and Development Program under Grant KMI (2018-03010).

Conflicts of Interest: The authors declare no conflict of interest.

Appendix A

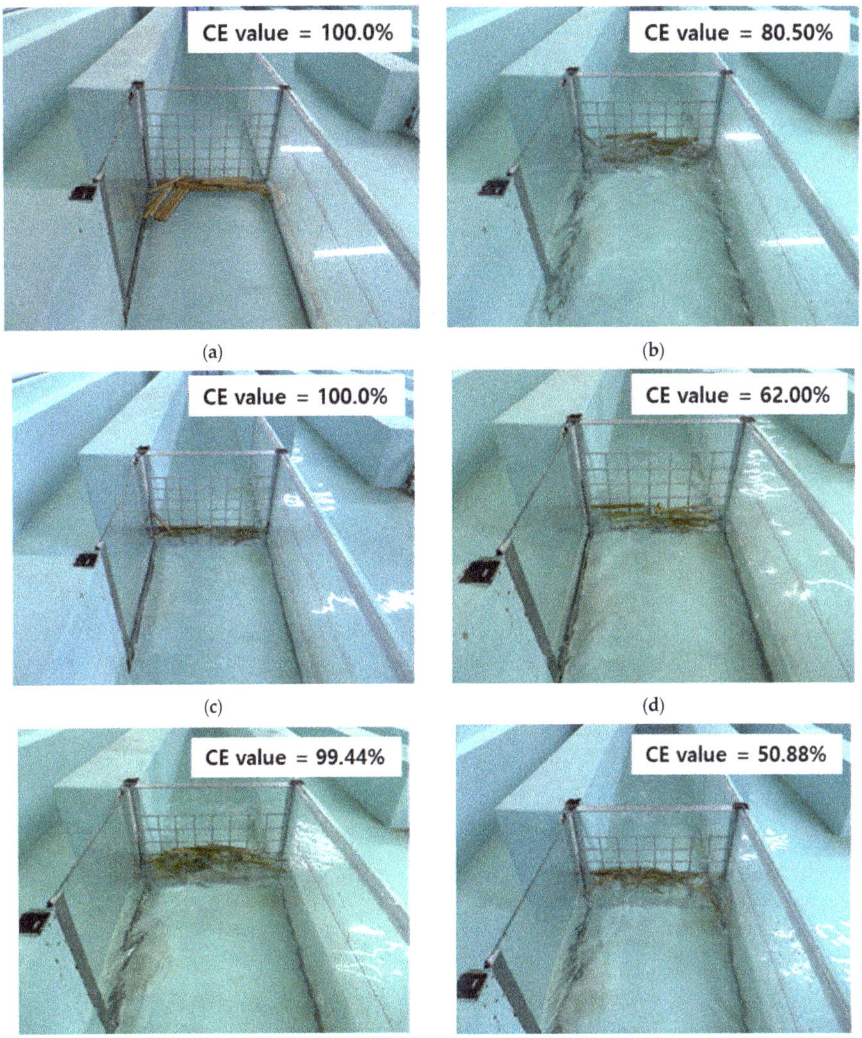

(a) (b) (c) (d) (e) (f)

Figure A1. *Cont.*

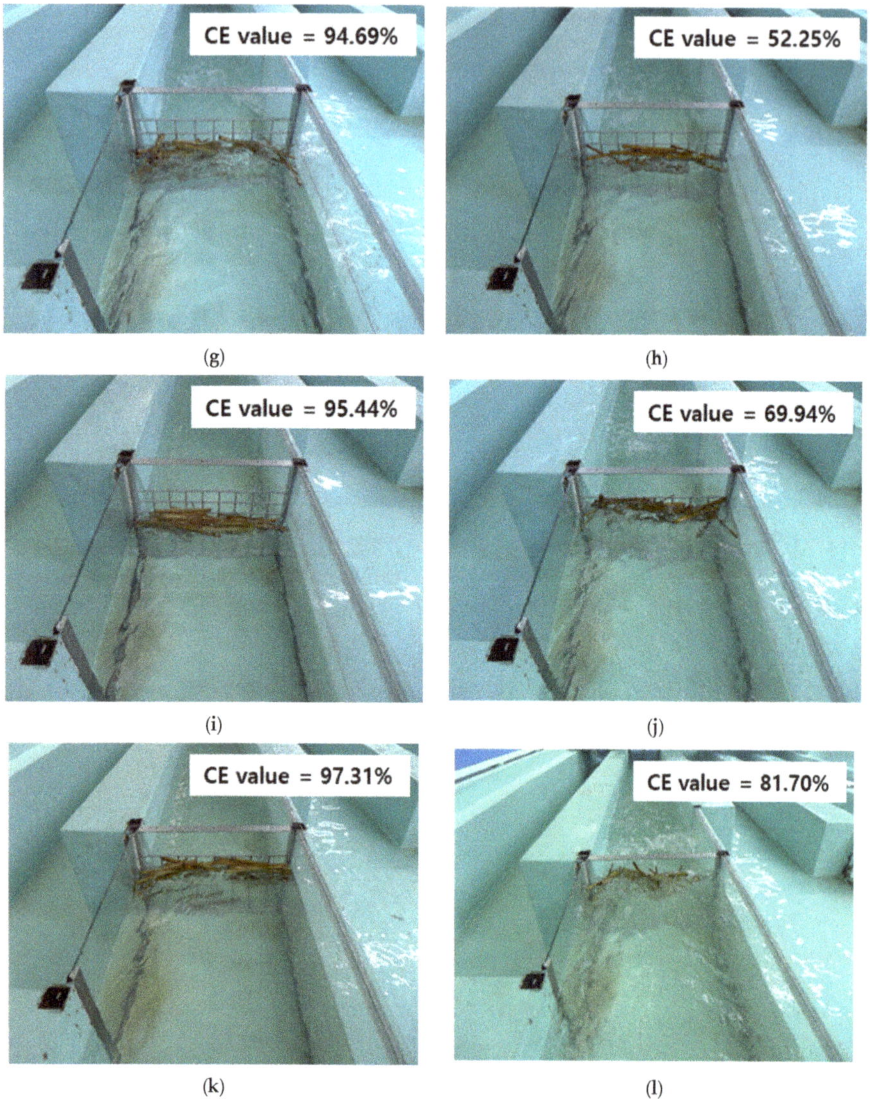

Figure A1. Capture efficiency of the driftwood capture trellis by experimental condition: (**a**) B400mm-H80mm-V0.3; (**b**) B400mm-H80mm-V1.8; (**c**) B400mm-H120mm-V0.3; (**d**) B400mm-H120mm-V1.8; (**e**) B400mm-H160mm-V0.3; (**f**) B400mm-H160mm-V1.8; (**g**) B400mm-H200mm-V0.3; (**h**) B400mm-H200mm-V1.8; (**i**) B400mm-H240mm-V0.3; (**j**) B400mm-H240mm-V1.5; (**k**) B400mm-H280mm-V0.3; (**l**) B400mm-H260mm-V1.2.

References

1. Chen, S.Y.; Xue, Z.C.; Li, M.; Zhu, X.P. Variable sets method for urban flood vulnerability assessment. *Sci. China Technol. Sci.* **2013**, *56*, 3129–3136. [CrossRef]
2. Zhu, Z.; Chen, Z.; Chen, X.; He, P. Approach for evaluating inundation risks in urban drainage systems. *Sci. Total Environ.* **2016**, *553*, 1–12. [CrossRef] [PubMed]

3. Yin, J.; Yu, D.; Yin, Z.; Liu, M.; He, Q. Evaluating the impact and risk of pluvial flash flood on intra-urban road network: A case study in the city center of Shanghai, China. *J. Hydrol.* **2016**, *537*, 138–145. [CrossRef]
4. Pachauri, R.K.; Allen, M.R.; Barros, V.R.; Broome, J.; Cramer, W.; Christ, R.; Church, J.A.; Clarke, L.; Dahe, Q.; Dasgupta, P.; et al. *Climate Change 2014: Synthesis Report. Contribution of Working Groups I, II and III to the Fifth Assessment Report of the Intergovernmental Panel on Climate Change*; Core Writing Team, Pachauri, R.K., Meyer, L.A., Eds.; IPCC: Geneva, Switzerland, 2014; p. 151.
5. Du, J.K.; Qian, L.; Rui, H.Y.; Zuo, T.H.; Zheng, D.P.; Xu, Y.P.; Xu, C.Y. Assessing the effects of urbanization on annual runoff and flood events using an integrated hydrological modeling system for Qinhuai River basin. *China. J. Hydrol.* **2012**, *464*, 127–139. [CrossRef]
6. Yin, J.; Yu, D.; Yin, Z.; Wang, J.; Xu, S. Modelling the anthropogenic impacts on fluvial flood risks in a coastal mega-city: A scenario-based case study in Shanghai, China. *Landsc. Urban Plan.* **2015**, *136*, 144–155. [CrossRef]
7. Ministry of Land, Transport and Maritime Affairs. *Urban Roadway Design Guideline*, 1st ed.; Ministry of Land, Transport and Maritime Affairs: Goyangsi, Korea, 2012.
8. Korea Expressway Corporation. *Design Criteria for the Roads in Mountain Area*, 1st ed.; Korea Expressway Corporation: Sejong-si, Korea, 2006.
9. Ministry of Land, Infrastructure and Transport. *Slope Drainage Design Guideline*, 1st ed.; Ministry of Land, Infrastructure and Transport: Jinjusi, Korea, 2016.
10. AASHTO. AASHTO Highway drainage guidelines. In *Task Force on Hydrology and Hydraulics*; American Association of State Highway and Transportation Officials: Washington, DC, USA, 1999.
11. Park, M.S.; Jo, J.H.; Yun, D.K.; Han, K.H. A Study on the improvement of rural drainage system to cope with climate change. In Proceedings of the Korea Water Resources Association Conference, Daegu, Korea, 20 May 2011; Korea Water Resources Association: Daegu, Korea, 2011.
12. Wong, T.S.W. Kinematic wave method for determination of road drainage inlet spacing. *Adv. Water Resour.* **1994**, *17*, 329–336. [CrossRef]
13. Wong, T.S.W.; Moh, W.H. Effect of maximum flood width on road drainage inlet spacing. *Water Sci. Technol.* **1997**, *36*, 241–246. [CrossRef]
14. Tu, M.-C.; Traver, R.G. Optimal Configuration of an Underdrain Delivery System for a Stormwater Infiltration Trench. *J. Irrig. Drain. Eng.* **2019**, *145*, 05019007. [CrossRef]
15. Zhang, H.; Wang, X.; Wang, L. An Analytical Solution of Partially Penetrating Hydraulic Fractures in a Box-Shaped Reservoir. *Math. Probl. Eng.* **2015**, *2015*, 726910. [CrossRef]
16. Burgi, P.H.; Gober, D.E. *Bicycle-Safe Grate Inlets Study: Hydraulic and Safety Characteristics of Three Selected Grate Inlets on Continuous Grades*, 1st ed.; Federal Highway Administration, U.S. Department of Transportation: Washington, DC, USA, 1977.
17. Pugh, C.A. *Bicycle-Safe Grate Inlets Study: Hydraulic Characteristics of Slotted Drain Inlets*, 4th ed.; Federal Highway Administration, U.S. Department of Transportation: Washington, DC, USA, 1980.
18. Brown, S.A.; Stein, S.M.; Warner, J.C. *Urban Drainage Design Manual: Hydraulic Engineering Circular No. 22*, 1st ed.; Federal Highway Administration, U.S. Department of Transportation: Washington, DC, USA, 1996.
19. Young-Il, K.; Jung-Cheol, B. Experimental investigation of effects of sediment concentration and bed slope on debris flow deposition in culvert. *J. Korean Soc. Civ. Eng.* **2011**, *31*, 467–474.
20. Shin, H.J.; Won, C.H.; Choi, Y.H.; Kim, T.Y.; Choi, J.D. Study of installation of sediment trap drain channel to reduce soil erosion from storm water runoff. *J. Korean Soc. Agric. Eng.* **2010**, *52*, 95–100.
21. Liu, J.; Nakatani, K.; Mizuyama, T. Effect assessment of debris flow mitigation works based on numerical simulation by using Kanako 2D. *Landslides* **2013**, *10*, 161–173. [CrossRef]
22. Hye-Jin, K.; Gyeong-su, J. Varied flow analysis for linear drainage channels, Korea Water Resources Association. *J. Korea Water Res. Assoc.* **2008**, *41*, 773–784.
23. Johnson, P.A.; Hey, R.D.; Horst, M.W.; Hess, A.J. Aggradation at Bridges. *J. Hydraul. Eng.* **2001**, *127*, 154–157. [CrossRef]
24. Schmocker, L.; Hager, W.H. Probability of Drift Blockage at Bridge Decks. *J. Hydraul. Eng.* **2011**, *137*, 470–479. [CrossRef]
25. Schmocker, L.; Hager, W.H. Scale Modeling of Wooden Debris Accumulation at a Debris Rack. *J. Hydraul. Eng.* **2013**, *139*, 827–836. [CrossRef]

26. Chin, D.A. Hydraulic analysis and design of pipe culverts: USGS versus FHWA. *J. Hydraul. Eng.* **2013**, *139*, 886–893. [CrossRef]
27. Dasika, B. New approach to design of culverts. *J. Irrig. Drain. Eng.* **1995**, *121*, 261–264. [CrossRef]
28. Hager, W.H. Generalized culvert design diagram. *J. Irrig. Drain. Eng.* **1998**, *124*, 271–274. [CrossRef]
29. Meselhe, E.A.; Hebert, K. Laboratory Measurements of Flow through Culverts. *J. Hydraul. Eng.* **2007**, *133*, 973–976. [CrossRef]
30. Guven, A.; Hassan, M.; Sabir, S. Experimental investigation on discharge coefficient for a combined broad crested weir-box culvert structure. *J. Hydrol.* **2013**, *500*, 97–103. [CrossRef]
31. Shrestha, B.B.; Nakagawa, H.; Kawaike, K.; Baba, Y.; Zhang, H. Driftwood deposition from debris flows at slit-check dams and fans. *Nat. Haz. Earth Sys. Sci.* **2012**, *61*, 577–602. [CrossRef]
32. Lim, Y.H.; Chun, K.W.; Kim, M.S.; Yeom, J.J.; Lee, I.H. Capture effect of slit dam for debris flow and woody debris with hydraulic model experiment -Focusing on A and D type. In *Proceedings of the 2008 Annual Summer Conference of the Korean Society of Forest Science, Jeonju-si, Korea, 21 August 2008*; Korean Society of Forest Science: Jeonju-si, Korea, 2008.
33. Geun-woo, J.; Young-hyeop, L.; Su-Yeon, N.; Su-Jin, J.; Yu-Seok, J.; Sang-yup, Y. Sediment and woody debris trap effect of h-type slit dam with model experiment. In *Proceedings of the 2010 Annual Summer Conference of the Korean Society of Forest Science, Seoul, Korea, 14 September 2010*; Korean Society of Forest Science: Seoul, Korea, 2010.
34. Jin-hak, K.; Geun-woo, J.; Jung-il, S.; Joo-woong, Y.; Se-myung, K.; Yong-rae, K. Effects of float-board screen for catching drift woods and debris flows in urban areas. In *Proceedings of the 2013 Conference of the Korean institute of Forest Recreation, Seoul, Korea, 11 April 2013*; Korean Institute of Forest Recreation: Seoul, Korea, 2013.
35. Douglas County. *Storm Drainage Design and Technical Criteria Manual*, 1st ed.; Douglas County: Englewood, CO, USA, 2008.
36. Kim, S.J.; Kang, J.G.; Kim, J.T. An experimental study on debris reduction system for culvert. *J. Korea Acad. Indust. Coop. Soc.* **2017**, *18*, 696–706.
37. Bradley, J.B.; Richards, D.L.; Bahner, C.D. *Debris Control Structures: Evaluation and Countermeasures*, 3rd ed.; U.S. Department of Transportation: Washington, DC, USA, 2005.
38. Ministy of Agriculture, Forestry and Flsheries. *Land Improvement Project Plan Design Standard Plan*, 1st ed.; Ministy of Agriculture, Forestry and Flsheries: Tokyo, Japan, 2019.
39. Ministry of Land, Infrastructure and Transport. *Road Design Manual of Mountain Area*, 1st ed.; Ministry of Land, Infrastructure and Transport: Sejong-si, Korea, 2007.
40. Town of Castle Rock. *Storm Drainage Design and Technical Criteria Manual*, 1st ed.; Town of Castle Rock: Castle Rock, CO, USA, 2019.
41. Department of Water Govement of Western Australia. *Stormwater Management Manual for Western Australia*, 1st ed.; Department of Water Govement of Western Australia: Perth, Australia, 2007.
42. National Institute for Land and Infrastructure Management. *Manual of Technical Standard for Designing Sabo Facilities against Debris Flow and Driftwood*, 1st ed.; National Institute for Land and Infrastructure Management: Sabo, Japan, 2016.
43. City of Toronto Technical Services Metro Hall. *Design Criteria for Sewers and Watermains*, 1st ed.; City of Toronto Technical Services Metro Hall: Toronto, ON, Canada, 2009.

applied sciences

MDPI

Article

Influence of the Ion Mass in the Radial to Orbital Transition in Weakly Collisional Low-Pressure Plasmas Using Cylindrical Langmuir Probes

Guillermo Fernando Regodón [1], Juan Manuel Díaz-Cabrera [2,*], José Ignacio Fernández Palop [1] and Jerónimo Ballesteros [1]

1 Departamento de Física, Campus Universitario de Rabanales, Universidad de Córdoba, 14071 Córdoba, Spain; z62rehag@uco.es (G.F.R.); fa1fepai@uco.es (J.I.F.P.); fa1bapaj@uco.es (J.B.)
2 Departamento de Ingeniería Eléctrica y Automática, Campus Universitario de Rabanales, Universidad de Córdoba, 14071 Córdoba, Spain
* Correspondence: el1dicaj@uco.es

Received: 15 July 2020; Accepted: 17 August 2020; Published: 19 August 2020

check for updates

Abstract: This paper presents an experimentally observed transition from the validity of the radial theories to the validity of the orbital theories that model the ion current collected by a cylindrical Langmuir probe immersed in low-pressure, low-temperature helium plasma when it is negatively biased with respect to the plasma potential, as a function of the positive ion-neutral collision mean free path to the Debye length ratio $\Lambda = \lambda_+/\lambda_D$. The study has been also conducted on argon and neon plasmas, which allows a comparison based on the mass of the ions, although no transition has been observed for these gases. As the radial or orbital behavior of the ions is essential to establish the validity of the different sheath theories, a theoretical analysis of such a transition not only as a function of the parameters Λ and $\beta = T_+/T_e$, T_+ and T_e being the positive ion and electron temperature, respectively, but also as a function of the ion mass is provided. This study allows us to recognize the importance of the mass of the ion as the parameter that explains the transition in helium plasmas. Motivated by these theoretical arguments, a novel set of measurements has been performed to study the relationship between the Λ and β parameters in the transition that demonstrate that the effect of the ion mean free path cannot be completely ignored and also that its influence on the ion current collected by the probe is less important than the effect of the ion temperature.

Keywords: plasma surface technology; cold plasma; ion temperature; ion-neutral charge-exchange collisions; ion mean free path; plasma diagnosis; Langmuir probe; sheath theories

1. Introduction

In low-pressure, low-temperature plasmas, the study of the positive ion current collected by the Langmuir probe is very important, as the smallness of the positive ion current collected by the Langmuir probe when it is polarized negatively with respect to the plasma potential allows local diagnosis of the plasma parameters with very low disturbance to the plasma. In other words, the positive ion sheath that is formed around the probe shields out the influence of the probe, which in plasmas with low plasma density is crucial [1–7]. On the other hand, many surface technological processes that use plasmas depend on the ion current that reaches the surface, and thus the control of the ion current is essential in this kind of technology. Among these processes, we have plasma-assisted chemical vapor deposition (PACVD), ion implantation, etching, surface coating, thin films, nanotechnology, etc. [8–13]. In the semiconductor industry, which is a major application of PACVD, the properties of the plasma must be closely examined in order to control the energy and the frequency of the ion impacts against the surface, so that the optimal conditions for ion implantation are obtained [8,10]. Therefore, both theoretical analysis and experimental studies of the ion sheath surrounding the surface to be treated are important.

The ion current collected by a Langmuir probe has been extensively studied from a theoretical point of view. There are two main groups of theories to explain the fall of the ions towards the probe. The orbital theories, of which orbital motion limited (OML) is the most frequently used, study the movement of the ions in orbits around the probe that are calculated using the applicable laws of conservation. Some ions have a trajectory that does not intersect with the probe surface and orbit back to the plasma, so that not all of the ions are collected by the probe [14–16]. The radial theories, the first being the Allen–Boyd–Reynolds (ABR) theory for a spherical Langmuir probe, study the plasma as a fluid, so that all the ions fall radially towards the probe and therefore all the ions that enter the sheath are collected by the probe [17]. The Allen–Boyd–Reynolds theory, which was soon adapted to cylindrical Langmuir probes by Chen [18], is valid for ions that have an ion temperature that is negligible when compared to the electron temperature, so that the parameter $\beta = T_+/T_e$, with T_+ and T_e as the positive ion and electron temperature, respectively, can be given a value of $\beta = 0$. The cylindrical radial model has been extended by the authors to $\beta \neq 0$ [1,2,19–24].

Both orbital and radial theories are used to diagnose plasmas to obtain plasma parameters such as plasma density. However, the values obtained using the two theories can be very different, the one predicted by the radial theory being up to an order of magnitude higher than the one predicted by the orbital theory. When the values of the plasma parameters deduced from these theories are compared to the values of the plasma parameters obtained using the much higher electron current in the electron saturation zone, depending on the plasma conditions, it is found that either the radial or the orbital theories are consistent with the well-established electron saturation zone theory. This implies a paradox when the ion current is used in plasma diagnosis, given that the appropriate theory that should be used is not known a priori before it is used in plasma diagnosis [1,4,25]. Actually, in many situations which depend on the plasma discharge power and the pressure, values for the ion current collected by the probe between the two theories are measured in experiments [5,26–28]. In two previous papers [26,28], the authors showed an experimentally observed transition in the positive ion current values that are derived from the radial theories and the orbital ones, as a function of the β parameter. The transition takes place only for helium plasmas. Similar experiments on argon and neon plasmas [1,2,19–24] show that the radial theory developed by the authors that takes into account the temperature of the ions, but does not take into account the mass of the ions or the ion mean free path, is successful in predicting the positive ion current.

Neither radial nor orbital models consider collisions in the calculation of the positive ion current of a negatively biased cylindrical Langmuir probe relative to the plasma. The most frequent collision of the ions in their fall towards the probe in low-pressure, low-temperature plasmas is the ion-neutral charge-exchange collision (INCEC) with the neutral atoms of the background gas [29,30]. In INCEC, an electron transitions from the neutral atom to the ion, effectively interchanging the moment between them. In radial models, this loss of moment is similar to a friction force in the fluid, so that the positive ion current is reduced. In the orbital models, the ions may lose their orbital velocity, so that, after the last collision, the ions will lose their angular momentum and fall radially towards the probe [1,2,5,7,17,18,26,28], increasing the positive ion current collected by the probe. The effect of collisions is opposed in both models and will affect not only surface technology but plasma diagnosis methods, both depending on the positive ion current.

Electropositive plasmas can be studied by means of three parameters—that is, the ion to electron temperature ratio, the ion mean free path to Debye length ratio and the ion mass [22,31]. The effect of the ion to electron temperature ratio was studied in previous works [4,22,24,26,28]. The INCEC mean free path is recognized as an important parameter to discriminate between the ABR and the OML behavior in recent theoretical works [29–32]. This article shows an experimentally observed transition from the ABR to the OML behavior as a function of the INCEC mean free path, λ_+, to the Debye length,

λ_D, ratio $\Lambda = \lambda_+ / \lambda_D$ [26,28]. This study allowed us to recognize the mass of the ion, m_+, as the critical parameter in the presence of the observed transition in helium plasmas. A theoretical justification of the transition as a function of the β, Λ and ion mass parameters is proposed.

In the Section 2, after this introduction, the experimental setup and measurement method are briefly cited. The Section 3 states the experimental measurement conditions and the methodology. The Section 4 presents the results. In the Section 5, supported by a set of novel measurements, a theoretical discussion about why the transition takes place only for the helium plasmas is proposed. Finally, the Section 6 is an exposition of the conclusions.

2. Experimental Setup and Measurement Method

A high-voltage DC discharge has been chosen for these experiments. The gas is introduced in a large Pyrex cylinder, 40 cm high and with an inner diameter of 31 cm, where two stainless steel electrodes are held 15 cm from each other, each electrode having a diameter of 8 cm. The anode is connected to the ground. The electrodes are supplied a high DC voltage by means of the low ripple/low noise-to-signal ratio KEPCO BHK 2000-0.1MG high-voltage DC power supply. The DC power supply is configured as a current supply, given that the discharge current is related to the electron density and it better serves to characterize the discharge compared to the DC voltage or the discharge power. The entering gas flow is controlled by a mass gas flow controller, MKS 247. A tungsten cylindrical, 6 mm long, 0.1 mm diameter Langmuir probe is placed in the diffuse afterglow of the plasma discharge. In this zone, the plasma is spatially homogeneous and the electron temperature is found to be the lowest, so that the effect of the ion temperature cannot be neglected and the ion to electron temperature ratio, β, is not negligible [1,2,4,7,26,28]. The neutral and positive ions are supposed to be thermalized. As the electrodes are very hot during the measurements, the ion and neutral atom common temperature is estimated to be 350 K [1,2,5,7,24,26,28,33,34].

The current-voltage characteristic of the cylindrical Langmuir probe can be used to obtain an indirect measurement of the parameters that characterize the plasma in the zone of the discharge where it is placed. The plasma potential, V_{plasma}, the floating potential, V_{float}, and the electron energy distribution function, EEDF, can be measured [1,2,4,6,26,34–41]. Regarding the measured EEDF, it is checked that, in every measurement, the EEDF can be considered as following the distribution of Maxwell–Boltzmann [1,4,6,26,28], which is essential since the assumption of a Maxwellian EEDF is made in both radial and orbital theories. The EEDF is used to perform the calculation of the electron temperature, T_e, and the electron density, n_e, which is equal to the ion density, n_+, by means of the quasi-neutrality condition, $n_e \approx n_+$. These values for n_e and T_e have been used as in all further calculations since they do not depend on the radial or orbital theory used to obtain the results. The discharge and the measurements are controlled, and the initial calculations are performed, using a LabView Virtual Instrument [1,4,6,7].

The experimental device has been designed with the objective of obtaining a low electron temperature plasma, so that the ion temperature, in the range of the ambient temperature, becomes non-negligible when compared to the electron one [1]. This property of the DC discharge allowed us to check the validity of the radial theory developed by the authors, which takes into account the temperature of the ions in argon and neon plasmas in the conditions of the discharge [1,2,19–24]. On the other hand, the size of the probe was chosen to be small enough so that the small-radius OML theory would be of applicability in the cases in which the orbital theories are applicable, as was found in the helium plasma in some conditions of the discharge. The LabView-controlled measuring system makes fast measurements, each taking only 4 ms, so that the temperature of the probe does not change during the measurement of the current-voltage characteristic. In order to make sure that the measurements are quick enough, the current-voltage characteristics were measured, starting both in the electron saturation zone and in the ion saturation zone, making no difference to the results of the measurements.

3. Experimental Measurement Conditions and Methodology

The high-voltage DC power supply has an upper limit of 2000 V and 100 mA. The discharge currents are typically much lower, so that the voltage limitation is the relevant one. For argon, neon and helium plasmas, the discharge DC current is always lower than 12.5 mA. The pressure range for the argon plasma is $p(\mathrm{Pa}) \in [2, 10]$; for the neon plasma, it is $p(\mathrm{Pa}) \in [10, 35]$, and for the helium plasma, it is $p(\mathrm{Pa}) \in [13, 37]$. Given that the transition in the validity of the orbital and radial theories is found in the helium plasma, a total of 448 current-voltage characteristics were measured, while in argon and neon plasmas, for which no transition was found, a total of 171 and 111 current-voltage characteristics, respectively, were measured. This set of measurements includes the measurements already published [26,28] together with additional measurements in which we explored the higher electron temperature range in the three kinds of plasmas, although no new insight was gained, as the measurements followed the same trend as the rest of the measurements, albeit extending the range in which the trend was observed. The measured electron density, n_e, gives values in the range from 9×10^{14} to 7×10^{15} m^{-3}, while the electron temperature, T_e, ranges from 1000 to 4400 K, corresponding to β values which vary from 0.08 to 0.35. The $\Lambda = \lambda_+ / \lambda_D$ parameter is obtained from the following expression.

$$\lambda_+ = \frac{1}{n_+ \sigma_{+-n}}, \tag{1}$$

with $\sigma_{+-\mathrm{n}}$ being the cross-section for positive ion-neutral collision [10,42], and $\lambda_D = \sqrt{\frac{\epsilon_0 k_B T_e}{e^2 n_e}}$ [10]. For each of the plasmas, the following apply.

- For argon plasmas, the cross-section for positive ion-neutral collision, under our experimental conditions, is $\sigma_{+-\mathrm{n}} = 7.84 \times 10^{-15}$ cm^2 [10,42]. Therefore, the ion mean free path for argon is the following:

$$\lambda_+ = \frac{133}{218 \cdot p(\mathrm{Pa})} (\mathrm{cm}). \tag{2}$$

  Therefore, the Λ values vary from 3.98 to 29.71 for the argon discharges studied in this article.
- For neon plasmas, the cross-section for the ion-neutral collision is $\sigma_{+-\mathrm{n}} = 4.25 \times 10^{-15}$ cm^2 [10,42]. Therefore, the ion mean free path for neon is the following:

$$\lambda_+ = \frac{133}{118 \cdot p(\mathrm{Pa})} (\mathrm{cm}). \tag{3}$$

  Therefore, the Λ values vary from 1.52 to 24.27 for the neon discharges studied in this article.
- For helium plasmas, the cross-section for the ion-neutral collision is $\sigma_{+-\mathrm{n}} = 3.99 \times 10^{-15}$ cm^2 [10,42]. Therefore, the ion mean free path for helium is the following:

$$\lambda_+ = \frac{133}{111 \cdot p(\mathrm{Pa})} (\mathrm{cm}). \tag{4}$$

  Therefore, the Λ values vary from 2.27 to 14.29 for the helium discharges studied in this article.

Note that the upper limit for the ion mean free path to Debye length ratio Λ for the three gases decreases with decreasing ion mass, such that for argon, the upper limit is $\Lambda_{max,Ar} = 29.31$, close to the double the helium Λ upper limit, $\Lambda_{max,He} = 14.29$. It is also interesting to compare these Λ ranges with the sheath edge, which has values always lower than $k \cdot \lambda_D$. The k value depends on the sheath edge criteria used, usually considered to be $4 < k < 8$ [20,34,36], but regardless of the criteria used, there is a range in which the ion mean free path is comparable to the size of the sheath.

Two novel series of measurements were performed in the helium plasma, one for constant background pressure $P = 20.2$ Pa and one for constant discharge current $I_d = 5.0$ mA.

The methodology that is followed in this work is based on the use of the Sonin plot [1,2,4,26,28], which uses the positive ion current per unit length collected by the probe, I_+, when it is biased at a fixed electric potential, V_p in k_BT_e/e units, e being the elementary charge and k_B the Boltzmann constant, referring to the plasma potential, V_{plasma} [43]. In the Sonin plot, the ordinate of the plot is the non-dimensional ion current,

$$y_{Sonin} = I'(y_{SP}) = \frac{I_+(y_{SP})}{er_pn_+}\sqrt{\frac{m_+}{2\pi k_BT_e}}, \tag{5}$$

where r_p is the probe radius and y_{SP} is defined as the non-dimensional probe potential $y_{SP} = -eV_p/k_BT_e$. The abscissa of the plot has the following expression:

$$x_{Sonin} = x_p{}^2I'(y_{SP}) = \frac{I_+(y_{SP})er_p}{\varepsilon_0}\sqrt{\frac{m_+}{2\pi k_B{}^3T_e{}^3}}, \tag{6}$$

which does not depend on the ion density, with x_p being the non-dimensional probe radius $x_p = r_p/\lambda_D$. The V_p value must be chosen carefully, so that it is negative enough to ensure that the current collected by the probe is almost exclusively positive ion current, and the electron current can be neglected. Regarding the other extreme, if the difference between the plasma potential and the probe potential is too high, the emission of secondary electrons from the probe would be accounted for as an increase in the positive ion current collected by the Langmuir probe [1,2,4,5,26,28]. As in other articles, we have chosen $y_{SP} = -eV_p/k_BT_e = 25$ [1,2,4,26,28], which accounts for V_p values in the range $V_p(\text{V}) \in [2.15, 9.48]$ for the given range of electron temperatures $T_e(\text{K}) \in [1000, 4400]$. A single point in the Sonin plot is obtained for each set of plasma conditions—that is, for each set of experimental values for $I_+(V_p)$, n_e and T_e [1,4,5,26,28]. The experimental Sonin plot point is placed in the Sonin plot and its position relative to the theoretical orbital and radial curves is analyzed. The plasma conditions are studied in relation to the position of the experimental Sonin plot points.

4. Results

The experimental points are plotted for the different plasmas in Figures 1–3. The experimental Sonin plot points are grouped in terms of the INCEC mean free path to Debye length ratio Λ, which allows a comparison between the three plasmas in which the scale of the sheath varies with the Debye length. We show the Sonin plot that includes the theoretical curves that correspond to the orbital and the radial theories in Figures 1–3. The complete solution by Laframboise [16] for the orbital theory is used, calculated from the fitting curves obtained by Peterson and Talbot [44]. The radial model developed by the authors for several β values, which converge to the Allen–Boyd–Reynolds model adaptation to cylindrical Langmuir probes by Chen for negligible ion temperature with respect to the electron temperature [19,21,22]. In Figure 1, we also show the experimental Sonin plot points corresponding to different ranges in the Λ parameter for argon plasmas. The results of the measurements show that the radial theory describes appropriately the positive ion current collected by a cylindrical Langmuir probe immersed in an argon plasma, in the conditions of the DC discharge used in the measurements.

Figure 2 shows the experimental Sonin plot points obtained from the different neon plasma discharge conditions measured. Figure 2 also includes the experimental points colored and symbolized as a function for the Λ parameter. As can be seen in Figures 1 and 2, the Λ parameter shows an evolution in the Sonin plot points—that is, the points for which the ion mean free path is longer are grouped to the right, while the points for which the ion mean free path is short are grouped to the left. These points are located over the radial model theoretical curves for $\beta \neq 0$ and fit well with the experimental β values for each point. Therefore, the influence of Λ is not very important in terms of the ions' behavior in this range for argon and neon plasmas.

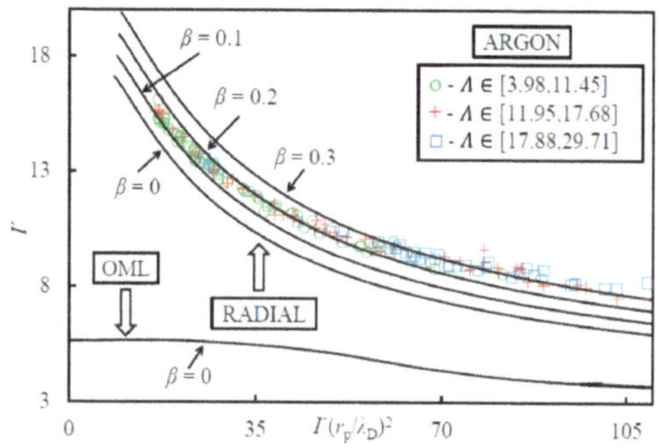

Figure 1. Argon plasma Sonin plot for the normalized probe potential $y_{SP} = 25$. Experimental data for the argon plasma: green circles for $3.98 \leq \Lambda \leq 11.45$, red crosses for $11.95 \leq \Lambda \leq 17.68$ and blue squares for $17.88 \leq \Lambda \leq 29.71$. In solid lines, the orbital and radial theoretical curves are shown.

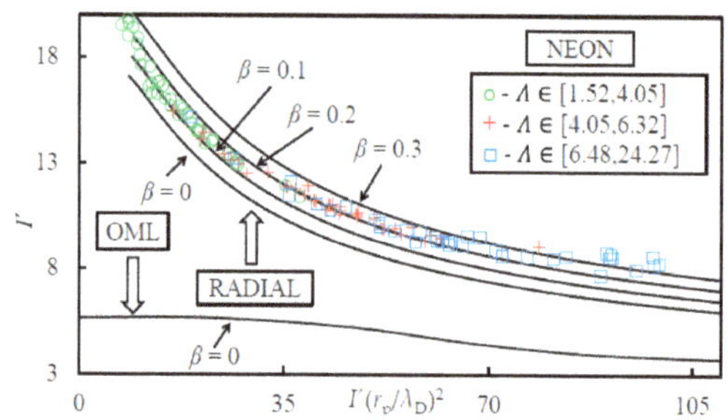

Figure 2. Neon plasma Sonin plot for the normalized probe potential $y_{SP} = 25$. Experimental data for the neon plasma: green circles for $1.52 \leq \Lambda \leq 4.05$, red crosses for $4.05 \leq \Lambda \leq 6.32$ and blue squares for $6.48 \leq \Lambda \leq 24.27$. In solid lines, the orbital and radial theoretical curves are shown.

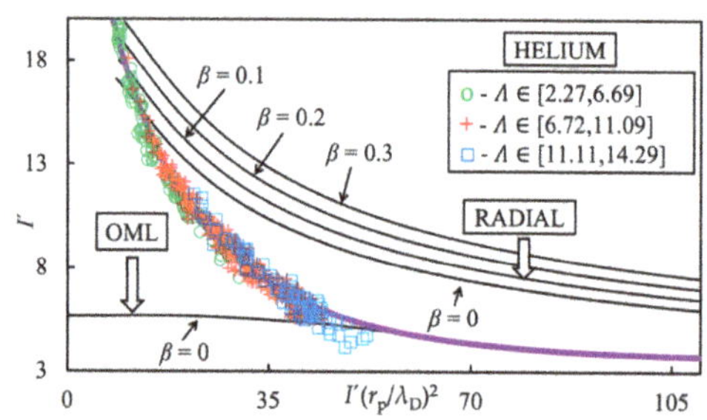

Figure 3. Helium plasma Sonin plot for the normalized probe potential $y_{SP} = 25$. Experimental data for the helium plasma: green circles for $2.27 \leq \Lambda \leq 6.69$, red crosses for $6.72 \leq \Lambda \leq 11.09$ and blue squares for $11.11 \leq \Lambda \leq 14.29$. In solid lines, the orbital and radial theoretical curves are shown. In purple bold line, the fitting curve for the experimental helium plasma Sonin plot points is shown.

As in Figures 1 and 2, Figure 3 shows the Sonin plot for the experimental points colored and symbolized as a function of the Λ parameter. On one hand, those points corresponding to the lower Λ values (green circles for $2.27 \leq \Lambda \leq 6.69$) are mainly placed in the radial zone or close to it, and none of them are placed in the orbital zone. On the other hand, those points corresponding to the higher Λ values (blue squares for $11.11 \leq \Lambda \leq 14.29$) are mainly placed in the orbital zone or close to it, and none of them are placed in the radial zone. Finally, those points corresponding to intermediate Λ values (red crosses for $6.72 \leq \Lambda \leq 11.09$) are mainly placed in the intermediate zone, where none of the theories are verified, and only a few of them are placed in the other zones. Nevertheless, Figures 1–3 show that the dependence on Λ does not seem to be so critical than that on β [26,28], since the Λ parameter depends not only on T_+ and T_e but also on $n_e \approx n_+$ and p through λ_+ and λ_D. Finally, Figure 3 also shows a fitting curve that may be used to diagnose this helium plasma in these conditions of pressure and discharge current, and that converges with OML in the range of abscissa of the Sonin plot, x_{Sonin}, of $[70, 100]$. However, although it is of great theoretical interest, the transition range of plasma conditions for the helium plasma should be avoided in plasma diagnosis by means of the Sonin plot, and the plasma conditions should be used in which one of the two theories, ABR or OML, is valid. The fitting curve, in purple bold line in Figure 3, follows the following formula:

$$y_{Sonin\ He,exp} = ae^{-b(x_{Sonin\ He,exp})^c} + d, \tag{7}$$

with $a = 32.11693$, $b = 0.13901$, $c = 0.77237$ and $d = 3.56418$.

5. Discussion

The finiteness of the ion mean free path and the experimental observation of its influence in the transition from radial to orbital behavior can be justified theoretically. After the last collision, the ions lose their orbital motion, so, for small β and Λ values, the OML theory cannot appropriately describe the ion current collected by the probe. For higher ion temperatures, after the last collision, the ions have a non-negligible velocity in a random direction. For the ions that have a direction that is predominantly in the azimuthal direction of cylindrical coordinates, the mean thermal velocity is important enough so that the trajectory of the ion will not intersect the probe and the ion will orbit around the probe back to the plasma. The mass of the ion thus becomes an essential parameter, since the mass of the ion is inversely proportional to the square of the thermal velocity.

$$\overline{v_{+,th}} = \sqrt{\frac{2k_BT_+}{m_+}}, \tag{8}$$

where m_+ is the mass of the ion. This fact explains why the transition is only found in the helium plasma but not in neon or argon plasmas.

The most frequent ion collision in this kind of plasma is INCEC, which removes an accelerated ion and results in a new ion with the temperature of the background gas [45,46]. It is interesting to note that the effect of INCEC in both radial and orbital models is the opposite: on one hand, in orbital models, usually the new ion created after an INCEC has a lower orbital kinetic energy, so the ion is more likely to be collected by the probe. If the orbital motion limited model is used, the exact potential profile of the sheath can be ignored, and the probability of the ion being collected can be calculated using the conservation laws. Therefore, the positive ion current collected by the probe is increased when INCEC is taken into account [47–49]. On the other hand, the effect of collisions in radial models is to reduce the mean velocity of the fluid particle, which is composed by many ions, in its fall towards the probe. That is, the positive ion current collected by the probe is reduced, having an effect similar to that of the transition from radial to orbital behavior. This effect can be estimated using the radial model that takes into account the ion temperature and the collisions of ions with neutral atoms developed by the authors [31] and proved to be opposite to the effect of the positive ion temperature in radial models, which increases the positive ion current [1,2,4,5,19,21,22,26,28]. It is interesting to note that, in neon and argon plasmas, the points in the Sonin plot shift up into regions of higher theoretical β values with increasing Λ, as can be seen in Figures 1 and 2, which is consistent with this development for radial theories—that is, less collisions in the ion sheath and a higher ion to electron temperature ratio have a similar effect in radial theories of increasing the current both in neon and argon plasmas, in the range of discharge parameters studied.

In the helium plasmas, the same increase in the ion mean free path to Debye length ratio Λ and in the ion to electron temperature ratio β has a very different influence on the movement of the ions in their fall towards the probe. The effect of collisions in the helium plasma can be studied with the available experimental data of the discharge. If the discharge current is decreased while maintaining the background pressure, the plasma and the electron density decrease [40,41]. Therefore, the Debye length increases, and the ion mean free path to Debye length ratio—that is, the Λ parameter—is decreased. We note in Equation (1) that the λ_+ parameter is constant if the background pressure is kept constant. Moreover, the measurements show that the electron temperature is higher when the discharge current is lower [40,41], so that the effect of a decrease in the discharge current is to cause a decrease in the β parameter value and a decrease in the Λ parameter (Figure 4). We have performed a series of novel measurements at constant background pressure, changing the discharge current, which show that the predicted trend is correct. This does not allow us to distinguish between the effect of the ion to electron temperature ratio β and the effect of the ion mean free path to Debye length ratio Λ in the experiments. However, if the discharge current is kept constant, the plasma density remains constant and so does the electron density [40]. Moreover, if the background pressure is decreased, the measurements show that the electron temperature, T_e, increases [40,41], causing a decrease in β. Accordingly, regarding the Λ parameter, we predict two opposite influences: (a) the Debye length increases with T_e, and thus the Λ parameter decreases; (b) the collisions are less frequent, increasing λ_+, and so the Λ parameter increases. Therefore, the trend that the Λ parameter will follow with changing background pressure alone cannot be predicted for a constant discharge current. We have also performed a series of novel measurements at constant discharge current to verify the influence of the background pressure on both the β and Λ parameters. Figure 5 shows that a decrease in the β parameter and an increase in the Λ parameter are experimentally found when the gas pressure decreases, according to argument (b) above. As the transition is found for decreasing background pressure, this proves that the most relevant parameter in the transition between radial and orbital behavior of the ions in the helium plasma is the positive ion to electron temperature ratio, β.

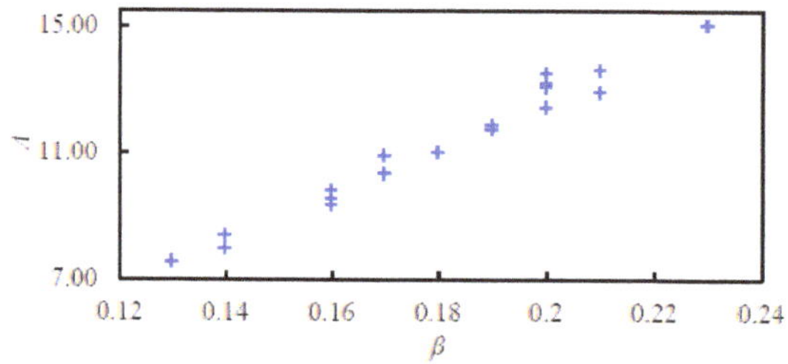

Figure 4. Increasing figure between β and Λ for constant background helium pressure $p = 20.2$ Pa.

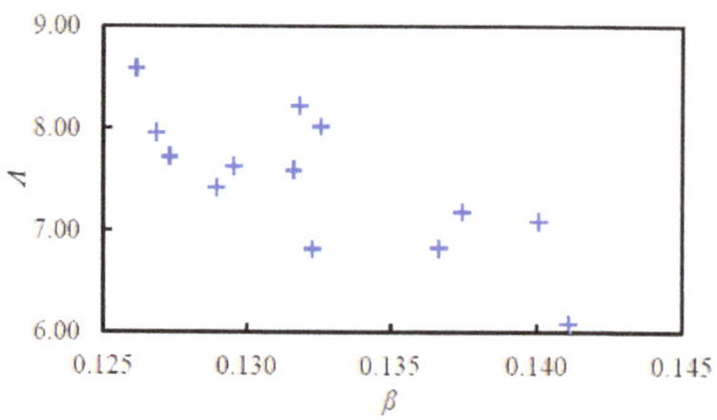

Figure 5. Decreasing figure between β and Λ for constant discharge current $I_d = 5.0$ mA in a helium discharge.

6. Conclusions

In plasma diagnosis experiments, when measuring the ion current collected by a cylindrical Langmuir probe, it is found that sometimes the orbital theories correctly predict the ion current and, at other times, the radial theories are found to be valid. Therefore, a transition from the validity of the orbital theories towards the validity of the radial theories is expected, depending on the experimental conditions of the plasma—that is, the ion temperature to electron temperature ratio, β, and the ion mean free path to Debye length ratio, Λ.

The transition has been theoretically justified in the context of INCEC, so that positive ions lose their translation kinetic energy in collisions with the neutral atoms of the background gas. Therefore, for small β and Λ values, after the last collision, the ions lose their orbital motion and the OML model is no longer valid to describe the ion current collected by the probe. This way, when collisions are included, the OML model provides higher positive ion current collected by the probe, i.e., approaching the values predicted by the ABR model. Alternatively, for higher Λ values, after the last collision, the ions are far away from the probe, and the orbital component of the ion thermal motion is high enough for those ions to fall towards the probe, following an orbital trajectory, diminishing the ion current collected by probe, i.e., approaching the OML model.

The aforementioned transition has been experimentally observed only for helium plasmas and not for argon and neon plasmas, and it has been justified due to the lower mass of the helium ions, which makes the helium thermal velocity higher for the same ion temperature. Therefore, only for higher Λ and β values in helium plasmas, the transition has been observed. Moreover, as an extreme case comparison, for electrons of even lower mass, the OML theory always predicts well the electron current collected when the probe is positively biased with respect to the plasma potential. Although it has been proven that the positive ion to electron temperature ratio, β, is a more relevant parameter than the INCEC mean free path to Debye length ratio Λ in the transition between radial and orbital behavior of the ions, the ion mass is crucial, since it determines the existence of the transition.

In view of these experimental results and theoretical arguments, the positive ion thermal motion and the collisions must be included in the sheath models, since they influence the radial or orbital behavior of the positive ion current to the substrate/probe, even though the collisions included in this article mainly take place outside of the sheath. This fact also explains that the β parameter has a more definite influence on the transition than the Λ one, as shown in Figures 4 and 5, but the influence of the ion mean free path to Debye length ratio Λ cannot be altogether ignored.

Author Contributions: Conceptualization, J.M.D.-C. and J.B.; methodology, J.M.D.-C. and J.B.; software, G.F.R. and J.M.D.-C.; validation, J.I.F.P. and J.B.; formal analysis, J.B.; investigation, G.F.R. and J.M.D.-C.; resources, J.I.F.P. and J.B.; data curation, G.F.R. and J.M.D.-C.; writing—original draft preparation, J.B.; writing—review and editing, G.F.R., J.M.D.-C., J.I.F.P. and J.B.; visualization, J.I.F.P. and J.B.; supervision, J.I.F.P. and J.B.; project administration, J.M.D.-C.; funding acquisition, G.F.R. and J.I.F.P. All authors have read and agreed to the published version of the manuscript.

Funding: This research has been partially funded by the Spanish Ministry of Science and Innovation, ref. no FIS2010-19951, which is partially financed with FEDER funds, and the FPU Program of the Spanish Ministry of Education (ref. FPU17/01387).

Conflicts of Interest: The authors declare no conflict of interest.

References

1. Díaz-Cabrera, J.M.; Fernández Palop, J.I.; Morales Crespo, R.; Hernández, M.A.; Tejero-Del-Caz, A.; Ballesteros, J. Virtual Instrument for automatic low temperature plasmas diagnostic considering finite positive ion temperature. *Measurement* **2014**, *55*, 66–73. [CrossRef]
2. Ballesteros, J.; Fernández Palop, J.I.; Hernández, M.A.; Morales Crespo, R. Influence of the positive ion temperature in cold plasma diagnosis. *Appl. Phys. Lett.* **2006**, *89*, 101501. [CrossRef]
3. Dengra, A.; Ballesteros, J.; Colomer, V.; Hernández, M.A. Investigation of the creation process in dc pulsed discharges. *J. Appl. Phys.* **1990**, *68*, 5507–5510. [CrossRef]

4. Díaz-Cabrera, J.M.; Lucena-Polonio, M.V.; Fernández Palop, J.I.; Morales Crespo, R.; Hernández, M.A.; Tejero-Del-Caz, A.; Ballesteros, J. Experimental study of the ion current to a cylindrical Langmuir probe taking into account a finite ion temperature. *J. Appl. Phys.* **2012**, *111*, 063303. [CrossRef]
5. Annaratone, B.M.; Allen, M.W.; E Allen, J. Ion currents to cylindrical Langmuir probes in RF plasmas. *J. Phys. D Appl. Phys.* **1992**, *25*, 417–424. [CrossRef]
6. Ballesteros, J.; Fernández Palop, J.I.; Hernaández, M.A.; Morales Crespo, R.; Del Pino, S.B. LabView virtual instrument for automatic plasma diagnostic. *Rev. Sci. Instruments* **2004**, *75*, 90–93. [CrossRef]
7. Lucena-Polonio, M.V.; Díaz-Cabrera, J.M.; Fernández Palop, J.I.; Morales Crespo, R.; Hernández, M.A.; Ballesteros, J. Mass spectrometry diagnosis of ion species in low-pressure plasmas. *Plasma Phys. Control. Fusion* **2011**, *53*, 124024. [CrossRef]
8. Chen, F.F. Industrial applications of low-temperature plasma physics. *Phys. Plasmas* **1995**, *2*, 2164–2175. [CrossRef]
9. Annemie, B.; Neyts, E.; Gijbels, R.; van der Mullen, J. Gas Discharge Plasmas and Their Applications. *Spectrochim. Acta Part B At. Spectrosc.* **2002**, *57*, 609–658.
10. Lieberman, M.A.; Lichtenberg, A.J. *Principles of Plasma Discharges and Materials Processing*; John Wiley & Sons: Hoboken, NJ, USA, 2005.
11. Shashurin, A.; Li, J.; Zhuang, T.; Keidar, M.; Beilis, I.I. Application of electrostatic Langmuir probe to atmospheric arc plasmas producing nanostructures. *Phys. Plasmas* **2011**, *18*, 073505. [CrossRef]
12. Weltmann, K.-D. Future trends in plasma science. *Plasma Process. Polym.* **2018**, *16*, e1890001. [CrossRef]
13. Laroussi, M. Cold Plasma in Medicine and Healthcare: The New Frontier in Low Temperature Plasma Applications. *Front. Phys.* **2020**, *8*. [CrossRef]
14. Mott-Smith, H.M.; Langmuir, I. The Theory of Collectors in Gaseous Discharges. *Phys. Rev.* **1926**, *28*, 727–763. [CrossRef]
15. Bernstein, I.B.; Rabinowitz, I.N. Theory of Electrostatic Probes in a Low-Density Plasma. *Phys. Fluids* **1959**, *2*, 112–121. [CrossRef]
16. LaFramboise, J.G. Theory of Spherical and Cylindrical Langmuir Probes in a Collisionless, Maxwellian Plasma at Rest. Ph.D. Thesis, University of Toronto, Toronto, ON, Canada, 1 June 1966.
17. E Allen, J.; Boyd, R.L.F.; Reynolds, P. The Collection of Positive Ions by a Probe Immersed in a Plasma. *Proc. Phys. Soc. Sect. B* **1957**, *70*, 297–304. [CrossRef]
18. Chen, F.F. Numerical computations for ion probe characteristics in a collisionless plasma. *J. Nucl. Energy. Part C Plasma Physics, Accel. Thermonucl. Res.* **1965**, *7*, 47–67. [CrossRef]
19. Fernández Palop, J.I.; Ballesteros, J.; Colomer, V.; Hernández, M.A. Theoretical ion current to cylindrical Langmuir probes for finite ion temperature values. *J. Phys. D Appl. Phys.* **1996**, *29*, 2832–2840. [CrossRef]
20. Morales Crespo, R.; Fernández Palop, J.I.; Hernaández, M.A.; Ballesteros, J. Analytical fit of the I–V characteristic for cylindrical and spherical Langmuir probes. *J. Appl. Phys.* **2003**, *94*, 4788. [CrossRef]
21. Morales Crespo, R. Analytical fit of the I–V probe characteristic for finite ion temperature values: Justification of the radial model applicability. *J. Appl. Phys.* **2004**, *95*, 2982. [CrossRef]
22. Regodón, G.F.; Fernández Palop, J.I.; Tejero-Del-Caz, A.; Díaz-Cabrera, J.M.; Carmona-Cabezas, R.; Ballesteros, J. Removal of singularity in radial Langmuir probe models for non-zero ion temperature. *Phys. Plasmas* **2017**, *24*, 103516. [CrossRef]
23. Regodón, G.F.; Fernández Palop, J.I.; Tejero-Del-Caz, A.; Díaz-Cabrera, J.M.; Carmona-Cabezas, R.; Ballesteros, J. Floating potential in electronegative plasmas for non-zero ion temperatures. *Plasma Sources Sci. Technol.* **2018**, *27*, 025014. [CrossRef]
24. Tejero-Del-Caz, A.; Fernández Palop, J.I.; Díaz-Cabrera, J.M.; Ballesteros, J. Radial-to-orbital motion transition in cylindrical Langmuir probes studied with particle-in-cell simulations. *Plasma Sources Sci. Technol.* **2015**, *25*, 01LT03. [CrossRef]
25. Pilling, L.S.; A Carnegie, D. Validating experimental and theoretical Langmuir probe analyses. *Plasma Sources Sci. Technol.* **2007**, *16*, 570–580. [CrossRef]
26. Díaz-Cabrera, J.M.; Ballesteros, J.; Fernández Palop, J.I.; Tejero-Del-Caz, A. Experimental radial motion to orbital motion transition in cylindrical Langmuir probes in low pressure plasmas. *Plasma Sources Sci. Technol.* **2015**, *24*, 25026. [CrossRef]
27. Tejero-Del-Caz, A.; Fernández Palop, J.I.; Díaz-Cabrera, J.M.; Regodón, G.F.; Carmona-Cabezas, R.; Ballesteros, J. Ion injection in electrostatic particle-in-cell simulations of the ion sheath. *J. Comput. Phys.* **2017**, *350*, 747–758. [CrossRef]

28. Díaz-Cabrera, J.M.; Fernández Palop, J.I.; Regodón, G.F.; Ballesteros, J. Accurate measurement of the ion saturation current collected by a cylindrical Langmuir probe in cold plasmas. *Plasma Process. Polym.* **2020**, 1–12. [CrossRef]
29. Franklin, R.N.; Snell, J. The plasma-sheath transition with a constant mean free path model and the applicability of the Bohm criterion. *Phys. Plasmas* **2001**, *8*, 643–647. [CrossRef]
30. Robertson, S. Sheaths in laboratory and space plasmas. *Plasma Phys. Control. Fusion* **2013**, *55*, 093001. [CrossRef]
31. Regodón, G.F.; Fernández Palop, J.I.; Díaz-Cabrera, J.M.; Ballesteros, J. Influence of collisions in a fluid model for the warm-ion sheath around a cylindrical Langmuir probe. *Plasma Sources Sci. Technol.* **2019**, *28*, 115017. [CrossRef]
32. Tsankov, T.V.; Czarnetzki, U. Information hidden in the velocity distribution of ions and the exact kinetic Bohm criterion. *Plasma Sources Sci. Technol.* **2017**, *26*, 055003. [CrossRef]
33. Gudmundsson, J.T.; Kimura, T.; A Lieberman, M. Experimental studies of O2/Ar plasma in a planar inductive discharge. *Plasma Sources Sci. Technol.* **1999**, *8*, 22–30. [CrossRef]
34. Godyak, V.A.; Piejak, R.B.; Alexandrovich, B.M. Probe diagnostics of non-Maxwellian plasmas. *J. Appl. Phys.* **1993**, *73*, 3657–3663. [CrossRef]
35. Regodón, G.F.; Fernández Palop, J.I.; Díaz-Cabrera, J.M.; Ballesteros, J. Floating potential calculation for a Langmuir probe in electronegative plasmas and experimental validation in a glow discharge. *Plasma Phys. Control. Fusion* **2019**, *61*, 095015. [CrossRef]
36. Swift, J.D.; Schwar, M.J.R. *Electrical Probes for Plasma Diagnosis*; Iliffe Books Ltd.: London, UK, 1970.
37. Demidov, V.I.; Ratynskaia, S.V.; Rypdal, K. Electric probes for plasmas: The link between theory and instrument. *Rev. Sci. Instruments* **2002**, *73*, 3409–3439. [CrossRef]
38. Curreli, D.; Chen, F.F. Equilibrium theory of cylindrical discharges with special application to helicons. *Phys. Plasmas* **2011**, *18*, 113501. [CrossRef]
39. Stamate, E. Status and challenges in electrical diagnostics of processing plasmas. *Surf. Coat. Technol.* **2014**, *260*, 401–410. [CrossRef]
40. Fernández Palop, J.I.; Ballesteros, J.; Colomer, V.; Hernaández, M.A. A new smoothing method for obtaining the electron energy distribution function in plasmas by the numerical differentiation of the I-V probe characteristic. *Rev. Sci. Instruments* **1995**, *66*, 4625–4636. [CrossRef]
41. Fernández Palop, J.I.; Ballesteros, J.; Colomer, V.; Hernaández, M.A. Transient processes in an Ar+I2 dc discharge. *J. Appl. Phys.* **1996**, *80*, 4282–4291. [CrossRef]
42. Maiorov, S.A.; Petrov, O.F.; Fortov, V.E. Calculation of resonant charge exchange cross-sections of ions Rubidium, Cesium, Mercury and noble gases. In Proceedings of the 34th EPS Conference on Plasma Physics, Warsaw, Poland, 2–6 July 2007.
43. Sonin, A.A. Free-molecule Langmuir probe and its use in flow-field studies. *AIAA J.* **1966**, *4*, 1588–1596. [CrossRef]
44. Peterson, E.W.; Talbot, L. Langmuir probe response in a turbulent plasma. *AIAA J.* **1970**, *8*, 1391–1398. [CrossRef]
45. Phelps, A.V. The application of scattering cross sections to ion flux models in discharge sheaths. *J. Appl. Phys.* **1994**, *76*, 747–753. [CrossRef]
46. Riemann, K.-U. Kinetic theory of the plasma sheath transition in a weakly ionized plasma. *Phys. Fluids* **1981**, *24*, 2163. [CrossRef]
47. Zakrzewski, Z.; Kopiczynski, T. Effect of collisions on positive ion collection by a cylindrical Langmuir probe. *Plasma Phys.* **1974**, *16*, 1195–1198. [CrossRef]
48. Sternovsky, Z.; Robertson, S.; Lampe, M. Ion collection by cylindrical probes in weakly collisional plasmas: Theory and experiment. *J. Appl. Phys.* **2003**, *94*, 1374. [CrossRef]
49. Sternovsky, Z.; Robertson, S.; Lampe, M. The contribution of charge exchange ions to cylindrical Langmuir probe current. *Phys. Plasmas* **2003**, *10*, 300–309. [CrossRef]

applied sciences

MDPI

Article

Performance and Modeling of a Two-Stage Light Gas Gun Driven by Gaseous Detonation

Weiqi Tang [1,2], Qiu Wang [3,*], Bingchen Wei [1,2,4,*], Jiwei Li [3], Jinping Li [3], Jiahao Shang [2,3], Kun Zhang [1] and Wei Zhao [2,3]

1 Key Laboratory of Microgravity (National Microgravity Laboratory), Institute of Mechanics, Chinese Academy of Sciences, Beijing 100190, China; tangweiqi@imech.ac.cn (W.T.); zhangkun@imech.ac.cn (K.Z.)
2 School of Engineering Science, University of Chinese Academy of Sciences, Beijing 100049, China; shangjiahao31@outlook.com (J.S.); zw@imech.ac.cn (W.Z.)
3 State Key Laboratory of High Temperature Gas Dynamics, Institute of Mechanics, Chinese Academy of Sciences, Beijing 100190, China; ljw@imech.ac.cn (J.L.); lijinping@imech.ac.cn (J.L.)
4 Center of Materials Science and Optoelectronics Engineering, University of Chinese Academy of Sciences, Beijing 100049, China
* Correspondence: wangqiu@imech.ac.cn (Q.W.); weibc@imech.ac.cn (B.W.)

Received: 14 May 2020; Accepted: 23 June 2020; Published: 25 June 2020

Abstract: A two-stage light gas gun driven by gaseous detonation was newly constructed, which can make up for the disadvantages of the insufficient driving capability of high-pressure gas and the constraints of gunpowder. The performance of the gas gun was investigated through experiments and a quasi-one-dimensional modeling of it was also developed and described in detail. The model accounts for the friction and heat transfer to the tube wall for gases by adding a source term. An improved model has been established to consider the inertial loads in the piston or projectile and model the friction force with the tube wall. Besides, the effects of pump tube pressure on the performance of the gas gun are also investigated numerically. Simulations of the pressure histories in the pump tube and the piston and projectile velocities were conducted. A good agreement was observed between the computational predictions and experimental results. The results showed that the friction between the piston and wall had only small influence on the piston velocity. The proposed numerical approach is suitable for the development of two-stage light gas guns and tests of the operating conditions.

Keywords: two-stage light gas gun; detonation; quasi-one-dimensional; piston; projectile

1. Introduction

A gasdynamic gun is a kind of test facility in which the test models or projectiles are launched at desired velocities and the aerodynamic properties of the flying models are measured during its flight, or shock and damage of the targets are measured upon the projectile impact [1]. With the invention of two-stage light gas gun by Crozier in 1946 and Hume in 1950s [2], especially hypervelocity ballistic range, was widely used in the fields of hypervelocity aerothermodynamics and hypervelocity impact effects as the results of development of hypervelocity reentry vehicles, protection of structures/shields against space debris, and kinetic energy weapons [3–8]. Although there are several other kinds of hypervelocity launchers, such as the electro-magnetic rail gun and coil gun, the two-stage light gas gun remains the most reliable and frequently used device in laboratories. Relative to other launchers, the two-stage light gas gun has advantages in considerable versatility, maturity of technology, low cost, and capability of launching test models of different shapes at hypervelocity.

Regardless of the driving method, the energy of the projectile depends on the energy that is stored in the driver section. Therefore, the driving method is of critical significance to the two-stage light gas gun. The simplest method is to utilize the high-pressure gas filled in driver section as the driver gas directly [9]. However, the driving capability of this method is not often sufficient, i.e., the available speed or energy of projectile is often not high enough to meet the experimental requirements within the range of operable pressure. This driving method is therefore mainly limited to situations requiring relatively low speed or energy for the projectile. To improve the driving capability, gunpowder is often used. Owing to the release of chemical energy, the driving capability of gunpowder is much more powerful than that of high-pressure gas. For example, the largest two-stage light gas gun in the world, the Arnold Engineering Development Complex (AEDC) Range-G, is capable of launching a projectile of 10 kg to speeds higher than 4 km/s [10]. Many studies of this driving method have been conducted, and relevant theories and numerical methods have been developed [6,11–13].

Although most large-scale two-stage light gas guns are currently driven by gunpowder, this driving method also has disadvantages. First, the transport, storage, and use of gunpowder are highly specialized activities and are restricted to strictly qualified organizations; therefore, the use of gunpowder is often out of reach for many research institutes. Second, it is a cumbersome task to clean the gunpowder residue after each run. Therefore, it is necessary to develop an alternative driving method that not only makes up the deficiency of driving capability of the high-pressure gas but also is more available and convenient compared to methods requiring gunpowder.

Gas guns driven by the combustion of flammable and explosive gases have been investigated as an alternative driving method. The combustion light gas gun that uses H_2, O_2, and H_e mixture gas as the driving gas was first developed by the UTRON Company [14]. It is capable of launching a projectile of 0.2 kg to speeds higher than 2.8 km/s. The China Aerodynamics Research and Development Center (CARDC) and McGill University also developed combustion two-stage light gas guns capable of a speed of 5.6 km/s [15]. Deng has developed an interior ballistics model of combustion light gas gun and found that the combustion process could be successfully controlled by adding the proper quantity of diluents. Pressure oscillation and average temperature of the chamber were also reduced [16–18]. However, the combustion light gas gun has difficulty in getting uniform and predictable ignition or muzzle velocities.

The detonation, a rapid and violent form of combustion, could serve as a high performance driver due to its rapid energy release in explosions. The detonation driver, which was first proposed by Bird [19] and has been applied to shock tunnels successfully [20], has also been suggested as the concept of "detonation products gun" [21]. The combustion products behind the detonation wave have much higher pressure and temperature than those of the initial state of the combustible mixture because of the strong compression produced by the leading shock wave with a Mach number range of 4 to 7. Subsequently, a single-stage gas gun driven by gaseous detonation was developed by Batchelor; the gas gun achieved velocities exceeding 3 km/s [22]. Li [23] discussed the internal ballistic process of a single-stage gas gun and verified its practical feasibility through experiments. Li also estimated the driving capability of a detonation driver for a two-stage light gas gun by solving the quasi-one-dimensional Euler equations; it was found that the detonation driver was sufficient to obtain projectile velocities of up to 8 km/s in a two-stage light gas gun. Unfortunately, experimental demonstrations and discussion of a two-stage gas gun driven by gaseous detonation are limited to date, nor its detailed numerical modeling. Obviously, numerical simulation is a simple method for the identification and assessment of new operating parameters of the test conditions without risk to the facility and hardware, as well as for the validation of new design concepts. It is also important for a detonation facility operating under the conditions of high temperature, high pressure, and presence of flammable and explosive gases.

In the present research, a two-stage light gas gun driven by gaseous detonation was newly constructed and is described in detail, including its piston and projectile measurements. A quasi-one-dimensional code is also developed to simulate the internal ballistics process. Viscous effects are incorporated by the well-known skin friction correlations for pipe flow. The friction force between the tube wall and the piston or projectile is considered by establishing an improved model that accounts for the inertial loads of the piston or projectile. Typical experiments are performed to validate the numerical method. Good agreements are obtained between the experimental and simulation results, including the piston/projectile velocity and pressure histories. The effect of parameters such as the pump tubes pressures on performance of the gun is also evaluated numerically.

2. Facility Description

2.1. Facility

This paper concerns the modeling and simulation of a particular newly built facility, namely the DBR30 two-stage light gas gun facility at the Institute of Mechanics Chinese Academy of Sciences (CAS). The facility consists of a detonation tube, a pump tube, a launch tube, and a target chamber, as shown in Figure 1. These components have inner diameters of 150 mm, 108 mm, 30 mm, and 1800 mm, respectively. The lengths of each section are shown in Figure 1.

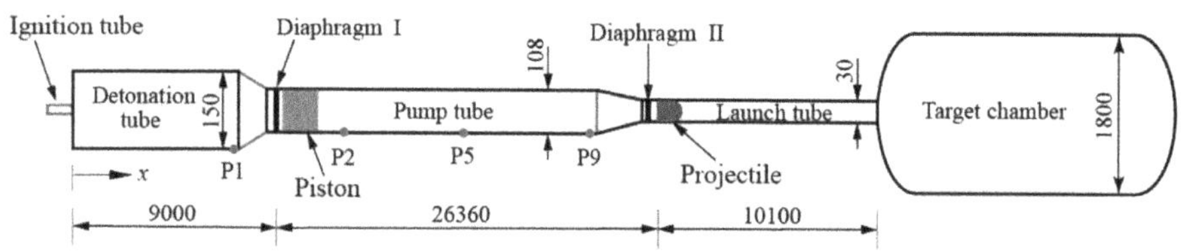

Figure 1. Schematic diagram of the two-stage light gas gun (not to scale, units in mm).

Diaphragm I separates the detonation tube and the pump tube, and diaphragm II separates the pump tube and the launch tube; both are scored steel diaphragm. The grooves in the form of a cross with different depths are designed to adjust the critical rupture pressure of the diaphragm and minimize fragmentation. Diaphragm II is designed to have a rupture pressure of 45 MPa or 75 MPa in the present study. An ignition tube is placed at the upstream end wall of the drive tube. The mixture in the ignition tube is initially ignited by electrical sparks. A high-temperature jet is formed and propagating into the driving section, thereby igniting the detonable gas directly. This will lead to a stable detonation wave traveling downstream in the detonation tube, behind which a Taylor wave follows and the products are at high temperature and pressure.

Piezoelectric pressure transducers (CY-YD-211, made by SINOCERA PIEZOTRONICS, INC (Yangzhou, Jiangsu, China)), with a resonant frequency of more than 100 kHz, are mounted on the tube sidewall to record the pressure histories. The signals from the sensors are acquired by a signal conditioner and are processed on a PC-based data acquisition system at a sampling rate of 2 MHz. The transducers are labeled as P1 to P9 in the present study. P1 is located in the detonation tube and the others are placed in the pump tube. The origin of the coordinate system is located on the left side of the facility, and x-axis is in the longitudinal direction of the device. Thus, the coordinates of P1–P9 are x_1 = 8.21 m, x_2 = 10.55 m, x_3 = 12.65 m, x_4 = 18.65 m, x_5 = 24.65 m, x_6 = 28.25 m, x_7 = 28.54 m, x_8 = 30.54 m, and x_9 = 30.83 m, respectively.

2.2. Piston and Projectile

The schematic drawing of the piston is shown in Figure 2. The front and rear part are made of polytetrafluoroethene (PTFE), which can prevent abrasion of the pump tube and minimize friction losses. The diameter of the piston at positions a, b, and c in Figure 2 is 108 + (0.1–0.2) mm at room temperature before the experiments. The diameter at position d is manufactured to be 108 + (0.5–0.9) mm, which can prevent gas leakage from the pump tube to the detonation tube and make full use of the compressed light gas. Although the piston will freeze before the experiments, the diameter at position d can still satisfy a close fit with the pump tube. To further minimize friction losses, the diameter of the piston is locally reduced as shown in Figure 2. The deformable front and rear parts of the piston have a weight of 4.114 kg and 0.774 kg, respectively, and they are replaced after each test. The PTFE surrounds a steel metallic weight, with a diameter of 105 mm. The length of the steel weight is selected in order to change the weight of the piston and influence the compression process. In the present study, a steel weight of 5.552 kg is selected. Thus, the total weight of the piston is 10.44 kg, with a total length of 680 mm.

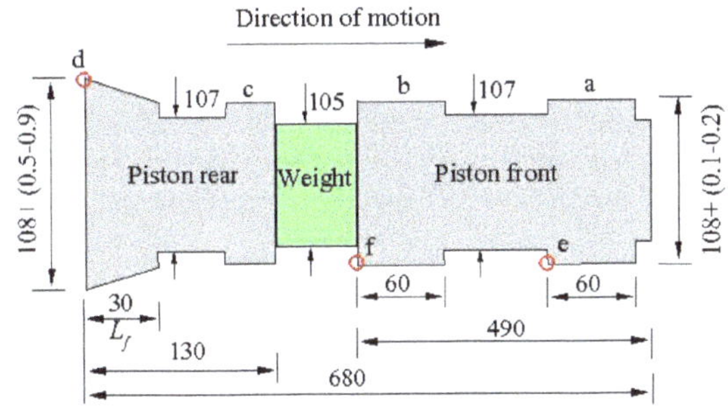

Figure 2. Schematic diagram of the piston (not to scale, units in mm).

Due to the geometric variation in the longitudinal direction as the piston moves past the piezoelectric transducers, the pressure around the sensors and the voltage of the piezoelectric transducers change. Typical pressure periods as the piston move past the sensors P3 and P8 are shown in Figure 3. Points d, e, and f in Figure 3 represent time when the corresponding point on the piston passes the sensor. In Figure 3a, the rising of the pressure at point d indicates the passing of the piston, as well as the arrival of the high pressure detonation gas. For the pressure sensors at the right section of the pump tube, they are already exposed to the high pressure light gas at the arrival of the piston. There are significant pressure changes due to the geometric variation in the longitudinal direction of the piston, as shown in Figure 3b. Thus, the pressure histories can be used to calculate the piston velocities at different positions.

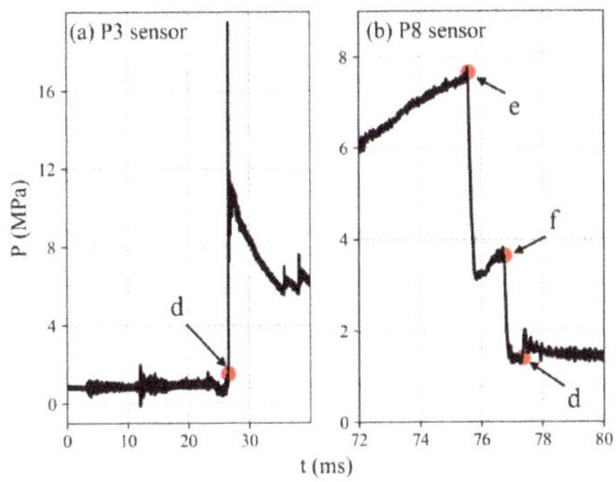

Figure 3. Pressure periods in the pump tube during the piston movement passing through the sensor. (**a**) P3 sensor; (**b**) P8 sensor.

Different shapes and weights of projectile can be used in the gas gun. However, only the muzzle velocity value and the measuring method are described here. A laser diagnostic system and high-speed photography were used to measure the projectile velocity. The laser diagnostic system can measure the projectile muzzle velocity directly, while the high-speed photography is an auxiliary method to verify the accuracy of the results of the laser diagnostic system results. The photography optical window is 3320 mm away from the launch tube exit. The snapshots in Figure 4 demonstrate the high-speed photography results processed by a script written in Python that incorporates edge detection and shape fitting methods. The velocity of the projectile is calculated using Equation (1):

$$v_{pro} = \frac{\Delta d}{\Delta t}\frac{R_{pro}}{R_{red}} \tag{1}$$

where Δd is the distance between the center of the red circle in two images, Δt is the time interval between two sequential frames, i.e., 13.3 μs in the present case; R_{pro} and R_{red} are the radius of the actual projectile and the radius of the red circle in the image, respectively. R_{red} and Δd have the unit of pixel and their values are related to the resolution of the high-speed camera. However, their values are detected from the images and have no influence to the actual projectile velocity calculation. The projectile in the present is a cylinder with a spherical head, as shown in Figure 4, and the diameter of the sphere is 30 mm.

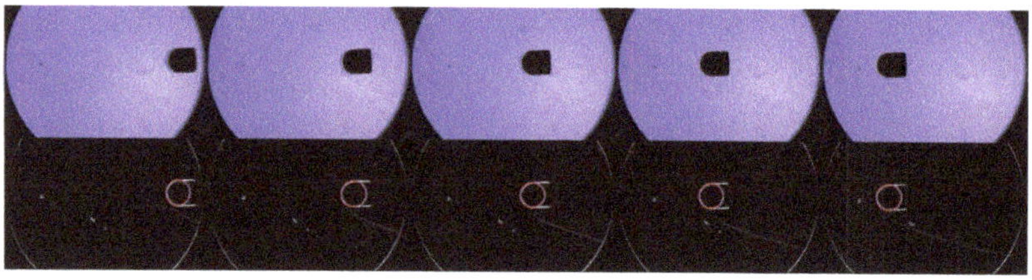

Figure 4. Results of high-speed photography with a time interval of 13.3 μs; the projectile has a diameter of 30 mm and the velocity is 2481 m/s; the pressure in the target chamber is 5000 Pa. The optical window was 3320 mm away from the muzzle exit.

3. Numerical Modeling

3.1. Governing Equations

Numerical simulations are necessary to understand the gas-dynamic processes and piston or projectile movement, and they are straightforward and provide detailed information. A simulation is also a simple method to identify and assess new operating parameters of the test conditions without risk to the facility and hardware, as well as for the validation of new design concepts. The two-stage gas gun driven by detonation involves detonation, flow of gas, motion of the piston and projectile, energy dissipation, and other complex chemical and physical phenomena. A two-dimensional axisymmetric investigation of the tube flows by solving the unsteady Navier-Stokes equation coupled with the heat conduction equation would be a more appropriate method. However, modeling all phenomena in detail is unrealistic in this method. For example, it is challenging to simulate the strong interactions between the detonation waves and the main diaphragm. In addition, the computer resources and simulation time are extensive, even for simple shock tube cases, not to mention the detonation tubes where complicated chemical reactions occur. Thus, a quasi-one-dimensional code is more attractive for the engineering design in the present study.

The flow of gas is modeled by the quasi-one-dimensional equations, and the detonation process in the drive gas is described by the two-step chemical reaction model modified by Sichel et al. [24]. The equations, written in conservation form, are

$$A\frac{\partial U}{\partial t} + \frac{\partial AF}{\partial x} - \frac{\partial A}{\partial x}H - S_c - AS_w = 0 \tag{2}$$

where the state vector $U = (\rho, \rho u, e, \rho\alpha, \rho\beta)^T$, the flux vector $F = (\rho u, \rho u^2 + p, (e + p)u, \rho\alpha u, \rho\beta u)^T$, the chemical reaction source term $S_c = (0, 0, 0, \dot{\omega}_\alpha, \dot{\omega}_\beta)^T$, the wall pressure source term $H = (0, p, 0, 0, 0)^T$, and the wall-friction and heat-transfer source term $S_w = (0, \tau, q, 0, 0)^T$; where ρ, u, e, p, and A are the density, velocity, total energy, and pressure of the gas, and the cross-sectional area, respectively. τ is the wall shear stress and q is the wall heat flux. α and β, which are associated with the two-step chemical reaction model describing the detonation process, are the process parameter of the chemical induction and the chemical transformation, respectively; $\dot{\omega}_\alpha$ and $\dot{\omega}_\beta$ are the rates of the chemical induction and the chemical transformation, respectively. For the details of this model, please refer to Ref. [24]. For the oxyhydrogen detonation, eight components, H_2, O_2, HO, H_2O, H_2O_2, HO_2, O, and H, are taken into account in the computations, and for the methane-oxygen detonation, CH_4, CO_2, and CO are added. Based on this chemical reaction model and the dispersion controlled dissipation scheme proposed by Jiang [25], a code has been developed and successfully applied to the simulations of detonation-driven shock tubes [26].

The situation of the two-stage gas gun differs from that of the shock tube due to the presence of the piston and projectile. To account for the effect of the piston or projectile on the gas flow, the motion of it is treated as the moving boundary conditions for the governing Equation (2). The acceleration of the piston or projectile is calculated from its mass and the applied force. Since the acceleration of the gas at the base and front of the piston or projectile is the same as the piston or projectile itself, and the disturbance resulting from the motion of the piston or projectile to the gas is isentropic if the displacement of the piston or projectile is small enough within each time step, the equations along characteristics and isentropic process still hold [11]. Thus, the sound speeds and pressures at these locations are obtained. The other state quantities are calculated in terms of the equations of state and of sound speed. Provided that the boundary condition is determined, Equation (2) can be solved as routinely as that for the detonation-driven shock tube.

The piston or projectile is assumed to have a fixed mass, length, and cross-section area. Their states are given by a flag whether they are constrained, its tail positions x_p, x_m, and velocities v_p, v_m. Subscript "p" represents the piston, and "m" represents the projectile throughout this paper. Since the process of the piston and projectile are described using the same method, only the governing differential equations are displayed here.

$$\frac{dv_p}{dt} = \frac{1}{m_p}\left(A_p(p_B - p_F) - F_f\right) \tag{3}$$

$$\frac{dx_p}{dt} = v_p \tag{4}$$

where p_B and p_F are the pressures acting on the back and front faces of the piston, respectively.

3.2. Gas-Phase: Friction and Heat Transfer Terms

Although the boundary layer along the tube wall is not completely modeled in the formulation of the gasdynamic equations, its effects are modeled by the addition of a wall shear stress in the momentum equation and heat transfer in the energy equation.

The wall-skin stress on the gas is given by: $\tau = -0.5 f \rho \pi u|u| D$. The skin friction coefficient for the pipe flow is then computed as [12]:

$$\begin{cases} f' = 0.049\left(\mathrm{Re}'_D\right)^{-0.2} & \mathrm{Re}'_D \geq 5507 \\ f' = 0.00875 & 5507 \geq \mathrm{Re}'_D \geq 1828 \\ f' = \dfrac{16}{\mathrm{Re}'_D} & 1828 \geq \mathrm{Re}'_D \end{cases} \tag{5}$$

$f = f'\frac{T}{T'}$ is evaluated at the reference temperature as follows: $T' = 0.9T + 0.03M^2T + 0.46T_w$, where T is the cell temperature and T_w is the specified wall temperature, M is the Mach number. The Reynolds number is calculated as: $\mathrm{Re}'_D = \frac{\rho' u D}{\mu'}$, where u is the cell velocity and D is the tube diameter of the cell. ρ' and μ' are evaluated at the reference conditions: $\rho' = \rho\frac{T}{T'}, \mu' = \mu'(T', \rho')$.

The heat transfer from the gas to the walls is given by: $q = h\pi D(T_w - T')$, where the heat transfer coefficient is $h = \rho C_p|u|S_t$ and the Stanton number is given by the modified Reynolds analogy for pipe flows [27]:

$$S_t = \frac{f}{8}\mathrm{Pr}^{-\frac{2}{3}} \tag{6}$$

In the treatment of the viscous gas mixture, Wilke's mixture rule is applied with Maxwell's Power Law to calculate the dynamic viscosity of the gas [28]:

$$\mu = \mu_0\left(\frac{T}{T_{ref}}\right)^n \quad n = 0.76\ (\text{air})\ \text{or}\ 0.83\ (\text{detonation gas}) \tag{7}$$

According to Ref. [27], the Prandtl number is given approximately as:

$$\mathrm{Pr} = \frac{20\gamma}{39\gamma - 15} \tag{8}$$

where γ is the ratio of the specific heat for gases.

3.3. Friction between Piston and Wall

The implementations in the L1d code assume that the friction force is simply proportional to the pressure exerted either at the back or the front of the piston, and that the piston is a perfectly elastic body [27]. However, this method is specified not from physical considerations, but to produce the best agreement between the calculated and experimental values of object velocity or gas pressure. An improved model with a physical basis has been implemented for the Longshot by accounting for the inertia loads within the piston [29]. Since different types of pistons are used in different facilities; modifications were made based on the Longshot method to obtain a better agreement in the present study. Details are described below.

The longitudinal stresses σ_x of a moving piston are obtained by considering successive static stresses (due to equal pressures applied on both ends of the piston) and dynamic stresses (due to the remaining pressure difference and the corresponding acceleration of the piston). The longitudinal stresses within a piston in arbitrary motion are then obtained by summing the static contribution and the dynamic one, as expressed by Equation (9).

$$\sigma_x = \min\left(p_b, p_f\right) + \frac{4F_{inertia}}{\pi D^2} \tag{9}$$

where D is the diameter of the piston, $F_{inertia} = m_p|a_p|$ are the piston inertia forces during the acceleration.

According to Hook's law generalized to cylindrical coordinates, the strain–stress relationship in an isotropic medium are:

$$\varepsilon_r = \frac{1}{E}(\sigma_r - \nu(\sigma_\theta + \sigma_x)) \tag{10}$$

$$\varepsilon_\theta = \frac{1}{E}(\sigma_\theta - \nu(\sigma_r + \sigma_x)) \tag{11}$$

where E is Young's modulus of the material, σ_x and ε_r are the radial stresses and strain, σ_θ and ε_θ are the azimuthal stresses and strain.

For the specific case where the piston is bounded by the tube, only axial and radial deformations are allowed. As long as the diameter of the piston is larger than that of the tube, differences in the diameter Δr exist between the piston and tube, and then the radial and angular strains are obtained as follows:

$$\varepsilon_r = \frac{\partial u_r}{\partial r} \approx \frac{\Delta r}{R_p} = \frac{2\Delta r}{D_p} \tag{12}$$

$$\varepsilon_\theta = \frac{u_r}{r} + \frac{1}{r}\frac{\partial u_\theta}{\partial \theta} \approx \frac{\Delta r}{R_p} = \frac{2\Delta r}{D_p} \quad (13)$$

where R_p is the radius of the piston, and D_p is the diameter.

Thus, the radial stress can be obtained from Equations (9)–(13). Besides, an upper limit for the local stresses σ_r may be associated to the maximum stresses allowable before yield would take place.

$$\sigma_r = \min\left(\sigma_{yield}, \frac{2E\Delta r}{D_p(1-\nu)} + \frac{\nu}{1-\nu}\left(\min(p_b, p_f) + \frac{4F_{inertia}}{\pi D_t^2}\right)\right) \quad (14)$$

The normal force F_N corresponding to the radial stresses applied on all friction surfaces of the piston is defined as:

$$F_N = \sigma_r\, \pi D_p L_f \quad (15)$$

where L_f is the length of the piston in contact with the tube. The friction force F_f is then assumed to be proportional to the normal force F_N:

$$F_f = \mu_f F_N \quad (16)$$

The piston consists of PTFE and its Poisson's ratio ν is 0.4; the sliding friction coefficient μ_f of steel-PTFE is in the range of 0.04–0.05. Young's modulus is about 0.5 GPa, and the yield stress is about 20 MPa [30].

At room temperature, the diameter of the piston is manufactured to be larger than that of the tube. However, freezing of the piston at 0 °C for at least four hours before the experiments is required to be able to insert the piston into the tube. Since time is needed to pump the tubes and fill them with the initial gas, the experiments are usually finished 30 to 50 minutes after the piston has been inserted. Figure 5 shows the changes in the piston diameter during the thawing process at room temperature. Positions a to d on the piston are illustrated in Figure 2. A positive Δr indicates a tight fit between the piston and pump tube, and a negative Δr means a loose fit. Positions a, b, and c still have negative Δr values after 60 minutes, while position d has positive values. Thus, only the rear end of the piston contacts with the pump tube; hence, L_f is set to 30 mm in the computation. Since the diameter of the rear part changes gradually, as shown in Figure 2, the diameter at its center position is used to deduce Δr. The diameter at position d is 108 + (0.5–0.9) mm. Thus, Δr is set to 0.25–0.45 mm in the calculation. It will become evident in the following discussion that these values do not have a significant influence on the modeling results.

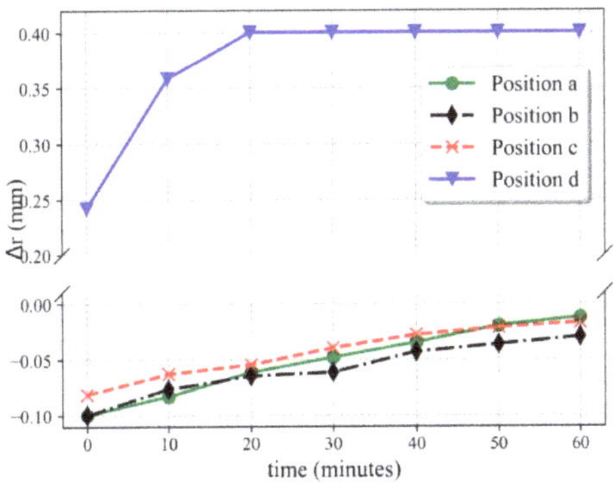

Figure 5. Changes in the radius of the piston at different positions versus time after removing it from the freezer.

4. Results and Discussion

Three particular test cases were selected from the available data, and the key parameters are listed in Table 1, including the projectile velocity from the experimental and computational fluid dynamics (CFD) results. Initial gas temperature was 293 K. However, in case 3, the system failed to capture the projectile velocity. The projectile velocity was measured at an optical window 3320 mm downstream of the launch tube exit. Since the pressure in the target chamber was usually set to a low pressure in the order of several kilo Pascal. It is assumed that the projectile speed has no significant attenuation.

Table 1. Operating conditions in the experiments and simulation.

Case		1	2	3
Detonation tube	Mixing molar ratio, H_2:O_2		3:1	
	Pressure (MPa)	1.5	2.0	2.7
Pump tube	Gas		H_2	
	Pressure (MPa)	0.8	0.8	0.45
	Piston mass (kg)		10.44	
	Rupture pressure (MPa)	45	45	75
Launch tube	Projectile mass (g)	19	100	31
Exp.	Projectile velocity (m/s)	2481	2260	
CFD		2839	2601	4854
	Relative error (%)	14.4	15.1	

4.1. Grid Independence Study

A grid convergence study was conducted using the above numerical method with three different grid resolutions (40,000, 60,000, and 80,000). The simulations were performed for Case 1. Figure 6 shows the velocity and pressure at the base of the piston over time for three grid resolutions. As shown in Figure 6, there are negligible differences for the piston base pressure and velocity between the three grid resolutions. Eventually, the resolution with 60000 grids was used in the present study.

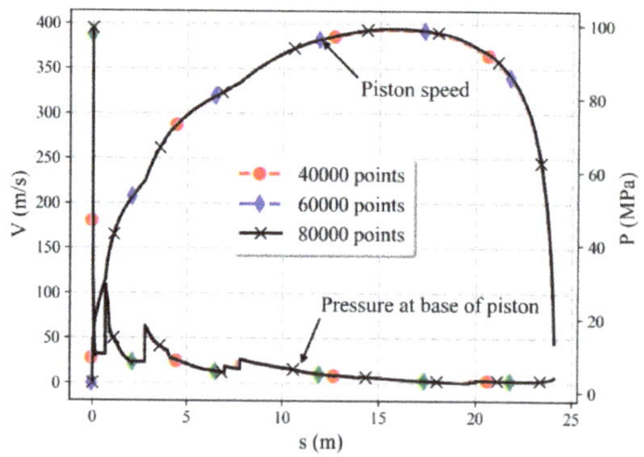

Figure 6. Speed and pressure at the base of the piston for Case 1 and three grid resolutions.

4.2. Comparison of Simulation and Experimental Results

Figure 7 shows the piston speed versus the travel distances for the three cases, including the experimental and numerical results. The numerical results agree well with the corresponding experimental ones using the above models; and the differences are within ± 10%, which is acceptable for the two-stage gas gun simulation. The numerical result without the piston-wall friction term for Case 1 is also shown in Figure 7. The results that include the piston friction are only 5% lower than those that do not include the piston friction. Thus, the values of L_f and Δr (defined in Section 3.3) have only small influences on the piston speed in the simulation.

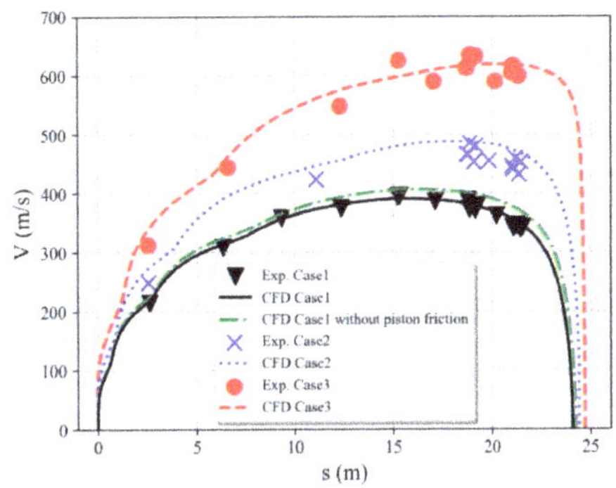

Figure 7. Piston speed versus piston travel distance.

Unlike a gas gun that is driven by high-pressure gases, which results in smooth changes in the piston speed, the detonation-drive gas gun produces fluctuations in the piston speed, as shown in Figure 7. This result is attributed to the complex interior ballistic process in the gas gun driven by detonation. The direct ignition at the upstream end in the detonation tube leads to a stable detonation wave traveling downstream. The detonation wave reflects from the base of the piston, generating a shock wave, and the motion of the piston generates compression waves. This shock wave then reflects several times between the end of the detonation tube and the base of the piston, which subjects the piston to a series of impulse loads, leading to fluctuations in the piston speed. The pressure at the base of the piston and projectile are shown in Figure 8; fluctuations also exist due to the shock wave reflections. The fluctuations are extensive in the initial stage and gradually weakened over time. However, it should be noted that the pressure at the base of the projectile was actually the pressure on diaphragm II before 82.8 ms. Since the projectile was placed close to diaphragm II, we use this pressure to monitor the pressure on diaphragm II or the pressure histories at the base of the projectile at the same time.

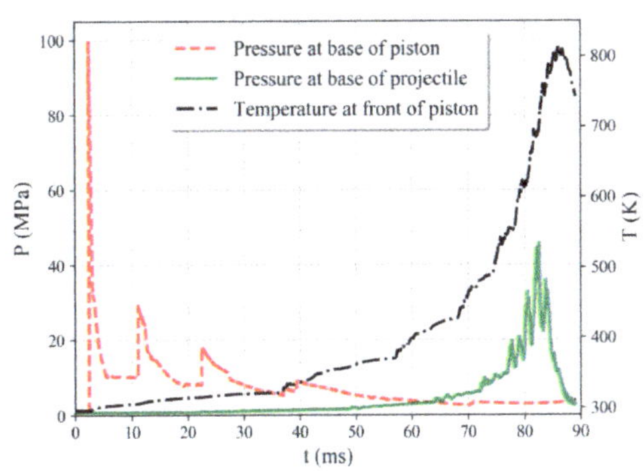

Figure 8. Simulated pressure and temperature at the base and front of the projectile or piston for Case 1.

In the simulation of Case 1, diaphragm II ruptures at 82.8 ms, and the projectile exits the launch tube at 88.2 ms; the corresponding times are 56.7 ms and 59.8 ms for Case 3, respectively. Thus, the pressure histories from 0 to 90 ms for Case 1 and 0 to 60 ms for Case 3 are chosen in the present discussion. Figures 9 and 10 show the experimentally measured and computed pressure histories at the monitoring points in the pump tube. In Figure 9, the pressure histories of P3 are displayed for Case 1 and Case 3. The first significant pressure increase was due to the passing of the piston or the arrival of the high-pressure detonation gas. The pressure changes following the increase indicate the complicated gas-dynamic processes between the piston base and the left wall of the facility. The simulation results show a similar trend as the experimental results. Additionally, the pressure value of the experiment agrees well with that of the simulation in the initial stage. However, the differences increase over time. It is believed that this result is attributable to a transducer problem rather than a code problem and is related to the influence of the high-temperature gas on the piezoelectric pressure transducers, which are supposed to operate at a temperature of 353 K. However, the temperature of the detonation gas is higher than 3000 K, and the temperature of the light gas also increases as it is compressed, as shown in Figure 8. The temperature of the light gas at the front of the piston exceeds 400 K at 60 ms for Case 1, and the maximum temperature exceeds 800 K. The temperatures are even higher for Case 2 and Case 3 because of higher piston speeds. Thus, the pressure transducer is exposed to high-temperature gases. Pressure differences occur and increase after a specific time. This phenomenon is also observed in Figure 10. More suitable high-temperature pressure sensors will be used in the future to validate the numerical results.

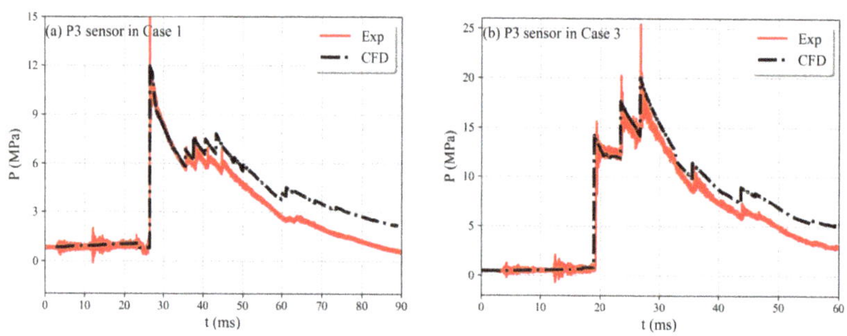

Figure 9. Pressure histories of the P3 sensor. (**a**) Case 1; (**b**) Case 3.

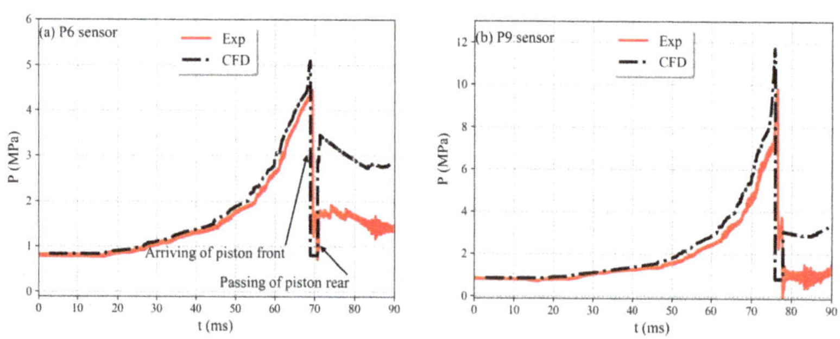

Figure 10. Pressure histories for Case 1. (**a**) P6 sensor; (**b**) P9 sensor.

The pressure changes due to the geometric variation in the longitudinal direction of the piston (see Figure 3b) are difficult to simulate. Thus, the pressure between the period of piston front and piston rear in the simulation is set equal to the initial pressure in the pump tube, i.e., 0.8 MPa for Case 1, as shown in Figure 10a between the time moment of 68.8 ms to 70.7 ms. However, they do not participate in the calculation and have no influence to the simulated results. Additionally, it can be found that there are larger discrepancy between CFD and Exp for P6 and P9 sensor after the passing of the piston. The reason may attribute to the inaccurate estimation of the energy exchange between the high temperature detonation gas (exceeds 3000 K) and the tube wall. This influence may go greater as time increases.

The launch tube has a relatively small diameter, and it would be complicated to measure the projectile velocity along the tube wall. Thus, only the muzzle velocity value is measured in the experiments; the results for Case 1 and 2 are listed in Table 1. However, in the simulation, we can obtain the projectile speed versus the projectile travel distance, as shown in Figure 11. The friction between the projectile and the wall is considered to be similar to that between the piston and the wall, as described in Section 3.3. The simulated projectile velocities at the exit of the launch tube are 2481 m/s, 2260 m/s, and 4854 m/s for Case 1, Case 2, and Case 3, respectively. The measured projectile velocities (Table 1) are 2481 m/s and 2260 m/s for Case 1 and Case 2, respectively. These values are about 14% lower than the numerical results. One reason for the difference is the rough estimate of the rupture pressure of diaphragm II. Scored diaphragms are used, and their rupture pressures are challenging to obtain experimentally under such a high pressure. The influence of the high-temperature light gas on the rupture pressure is also difficult to determine. Thus, only a semi-empirical formula is used to estimate the rupture pressure. The rupture pressure is 45 MPa for Case 1 and Case 2 and 75 MPa for Case 3. The influence of the rupture pressure on the projectile velocity is shown in Figure 12. Rupture pressures of 30 MPa, 45 MPa, and 75 MPa are considered for diaphragm II, and the other parameters are the same as in Case 1. A higher rupture pressure delays the rupture of diaphragm II, resulting in higher gas compression of the light gas; hence, the light gas provides a higher driving capability. The muzzle projectile velocities are 2553 m/s, 2601 m/s, and 2742 m/s for the three cases, respectively. In a future study, we will use simulations to model the rupture process of the diaphragm and improve the present method.

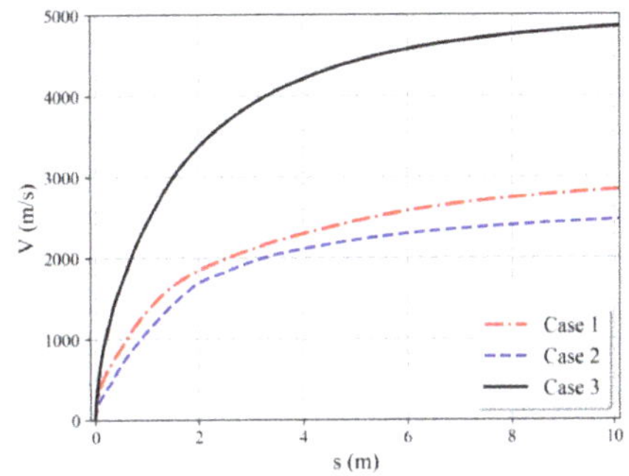

Figure 11. Simulated projectile velocity versus projectile travel distance.

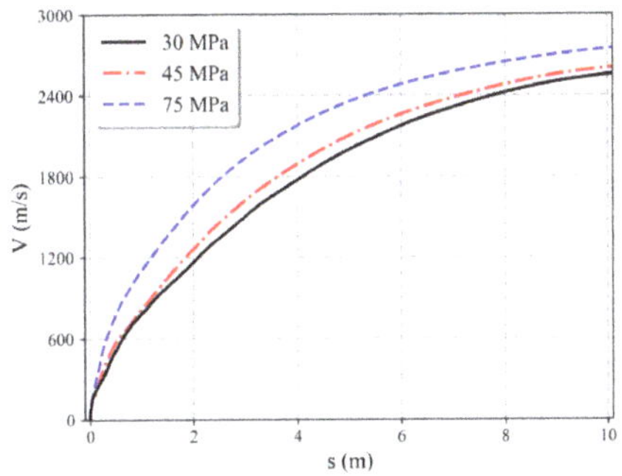

Figure 12. Simulated projectile velocities versus projectile travel distance for different rupture pressure values of diaphragm II.

Other reasons for the differences may be the roughness of the launch tube wall, the melting of the landing surface of the projective, improper data on the high strain rate deflection and yielding behavior of the projectile material, and that the viscous dissipation term between the high temperature light gas and tube wall are not completely accurate. However, the present simulation is acceptable for the identification and assessment of new operating parameters of the test conditions.

4.3. Effects of the Pump Tube Pressure

For any particular gun with a fixed geometry, the operating parameters or loading condition that can be adjusted to achieve the desired velocity include the initial gas conditions in the pump tube, the piston mass, the rupture pressure of diaphragm II, and the projectile mass. However, it is desirable to minimize the maximum base pressures to obtain the most uniform base pressure history and maintain gas reservoir pressures that are within the strength capabilities of the gun. Generally, the initial pressure of the pump tube and detonation tube is the easiest parameter to adjust.

The influence of the initial pump tube pressure on the gas gun was investigated; the operating conditions for Case 4 are listed in Table 2. Figure 13 shows the piston and projectile speed as a function of the pump tube pressure. Since the pump tube pressure tends to limit the acceleration of the piston, the piston speed decreases as the pump tube pressure increases at different detonation pressures. However, the pump tube pressure has little influence on the maximum piston speed, and the detonation pressure is the dominant influence. The projectile velocity has an optimal value at different pump tube pressures. For example, at a detonation pressure of 3.0 MPa, the maximum projectile speed occurs at a pressure of 0.4 MPa. The projectile base pressure in the launch tube is presented in Figure 14. When the initial pump tube pressure is lower than 0.4 MPa, the projectile has a relatively high base pressure of more than 130 MPa, indicating high acceleration in the early stages. However, the density decreases rapidly due to the lower initial pressure after the projectile has traveled about 2 m in the launch tube, leading to a significant decrease in the base pressure of the projectile. When the initial pump tube pressure is higher than 0.4 MPa, the peaks of the pressure of the projectile are much lower, and the curves are smooth due to stable acceleration. For the other different detonation pressures, the maximum projectile speed occurs at different initial pump tube pressure. At a detonation pressure of 1.5 MPa, an optimal projectile speed is not observed at the pump tube pressure range of 0.2 to 1.2 MPa.

Table 2. Operating conditions in the simulation of Case 4 and Case 5.

Case		4	5
Detonation tube	Mixing molar ratio, H_2:O_2	3:1	
	Pressure (MPa)	1.5, 2.0, 3.0, 5.0	5.0
Pump tube	Gas	H_2	
	Pressure (MPa)	0.2, 0.4, 0.6, 0.8, 1, 1.2	0.6
	Piston mass (kg)	10	12.5
	Rupture pressure (MPa)	45	150
Launch tube	Projectile mass (g)	19	10

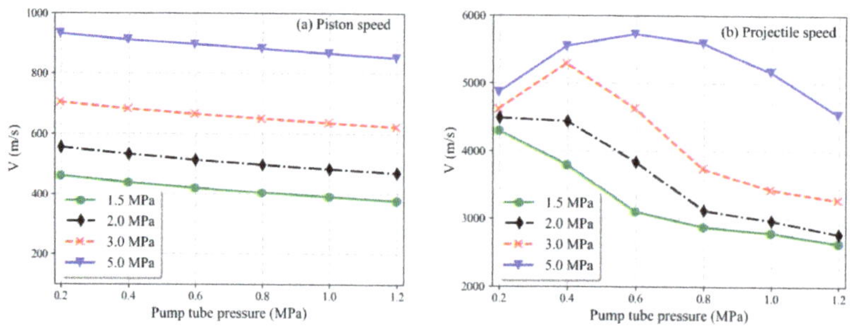

Figure 13. Simulated results at different detonation pressures. (**a**) maximum piston speed versus the initial pump pressure; (**b**) muzzle projectile speed versus the initial pump pressure.

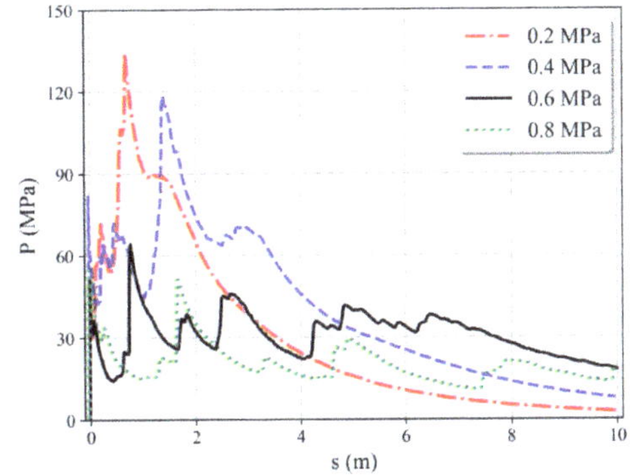

Figure 14. Simulated projectile base pressure versus the projectile travel distance for different pump tube pressure at a detonation pressure of 3.0 MPa.

We estimated the maximum capabilities of the gaseous detonation-driven two-stage light-gas gun. We increased the rupture pressure of diaphragm II to 150 MPa and maintained the maximum piston velocity at less than 900 m/s to protect the high-pressure section of the barrel. An appropriate set of calculation parameters was chosen (Case 5). The projectile velocity and base pressure are shown in Figure 15. The projectile muzzle velocity is more than 7100 m/s, and the maximum base pressure is close to 250 MPa, which is lower than the strength of the metal. Therefore, if we further optimize the launch parameters, the muzzle velocity of the projectile with a mass of less than 20 g can reach 6–8 km/s.

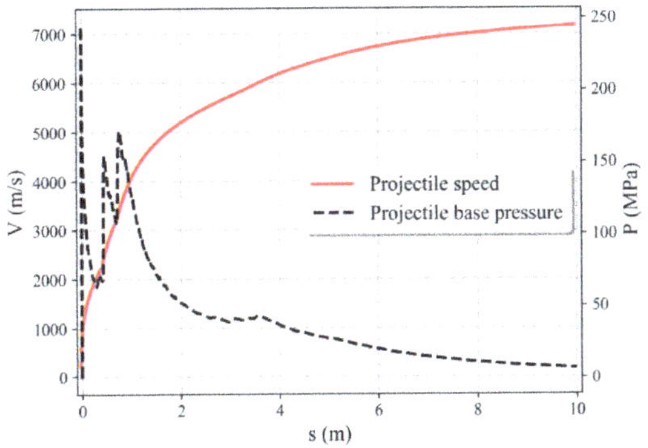

Figure 15. Projectile speed and base pressure for Case 5.

5. Conclusions

Due to strict controls of gunpowder, the use of traditional two-stage light-gas guns has limitations. Therefore, it is necessary to develop a new driving method for two-stage light-gas guns in laboratory conditions. In the present study, a two-stage light-gas gun driven by gaseous-detonation was developed, and its performance was evaluated. An accurate and rapid computational method was also implemented to simulate the flow or piston/projectile movement in the light-gas gun. The proposed method explains the wave propagation process in the tubes and provides a detailed piston/projectile trajectory. This type of model is required because different tests are conducted frequently in experimental facilities, and simulations are needed. The proposed quasi-one-dimensional model provides accurate simulation results of the internal ballistic process of the light-gas gun.

The results under different test conditions show that the quasi-one-dimensional modeling of the two-stage light-gas gun driven by gaseous detonation provides reliable performance data. The maximum muzzle velocity of the current design is 6–8 km/s. Although this study has focused on the DBR30 gas gun, the modeling is generic and may be applied to other facilities.

Author Contributions: Conceptualization, Q.W.; Data curation, J.S.; Formal analysis, J.L. (Jiwei Li) and J.L. (Jinping Li); Funding acquisition, K.Z.; Methodology, B.W. and W.Z.; Writing—original draft, W.T. All authors have read and agreed to the published version of the manuscript.

Funding: This research was funded by the Strategic Priority Research Program of the Chinese Academy of Sciences (Grant No. XDB22040303) and the National Natural Science Foundation of China (Grant No. 1197021407).

Acknowledgments: We thank Xijun Shi for his support in the experimental work.

Conflicts of Interest: The authors declare no conflict of interest.

References

1. Seiler, F.; Igra, O. (Eds.) *Hypervelocity Launchers*; Springer: Berlin, Germany, 2016; Volume 10, pp. 23–52.
2. Crozier, W.D.; Hume, W. High-Velocity, Light-Gas Gun. *J. Appl. Phys.* **1957**, *28*, 892–894. [CrossRef]
3. Canning, T.N.; Seiff, A.; James, C.S. *Ballistic-Range Technology*; AGARD: Neuilly sur Seine, France, 1970.
4. Lukasiewicz, J. *Experimental Methods of Hypersonics*; M. Dekker: New York, NY, USA, 1973; Volume 3.
5. Reda, D.C. Correlation of nosetip boundary-layer transition data measured in ballistics-range experiments. *AIAA J.* **1981**, *19*, 329–339. [CrossRef]
6. Strawa, A.W.; Chapman, G.T.; Canning, T.N.; Arnold, J.O. Ballistic range and aerothermodynamic testing. *J. Aircr.* **1991**, *28*, 443–449. [CrossRef]
7. Chapman, G. The ballistic range—Its role and future in aerothermodynamic testing. In Proceedings of the 17th Aerospace Ground Testing Conference, Nashville, TN, USA, 6–8 July 1992; American Institute of Aeronautics and Astronautics: Reston, VA, USA, 1992; p. 3996.
8. Marren, D.; Lu, F. *Advanced Hypersonic Test Facilities*; American Institute of Aeronautics and Astronautics: Reston, VA, USA, 2002; p. 138.
9. Junde, L. An analysis of launching parameters for a two stage light gas gun not driven by powder. *Explos. Shock Waves* **1995**, *3*, 229–240.
10. Carver, D.; Campbell, L.L.; Roebuck, B. Large-scale, hypervelocity, high-fidelity interceptor lethality development in AEDC's range G. *Int. J. Impact Eng.* **2008**, *35*, 1459–1464. [CrossRef]
11. Charters, A.C. Development of the high-velocity gas-dynamics gun. *Int. J. Impact Eng.* **1987**, *5*, 181–203. [CrossRef]
12. Bogdanoff, D.W.; Miller, R.J. *New Higher-Order Godunov Code for Modelling Performance of Two-Stage Light Gas Guns*; NASA's Ames Research Center: Mountain View, CA, USA, 1995.
13. Liang, S.C.; Huang, J.; Li, Y.; Jian, H.X. Numerical research on interior ballistics of two-stage light-gas gun. *Phys. Gases* **2012**, *7*, 80–84.
14. Kruczynski, D.; Massey, D.; Milligan, R.; Vigil, E.; Landers, B.; Meneguzzi, M. *Combustion Light Gas Gun Technology Demonstration*; UTRON Inc.: Manassas, VA, USA, 2007.
15. Dong, S.; Meng, C.; Xiao, Y.; Mo, J.; Zhang, M.; Wang, X.; Shi, S. Preliminary Study of Two-Stage Light Gas Gun Using Reactive Gas as Driving Energy. *Chin. J. High. Press. Phys.* **2017**, *31*, 182–186.
16. Deng, F.; Zhang, X.Y. Simulation of Hydrogen-Oxygen Combustion of Combustion Light Gas Gun Using Detailed Chemical Kinetics Model. *Acta Armamentarii* **2014**, *19*, 415–420.
17. Deng, F.; Zhang, X.Y.; Liu, N. Influences of ignition process and initial conditions on interior ballistic characteristics of combustion light gas gun. *Explos. Shock Waves* **2013**, *33*, 551–555.
18. Deng, F.; Liu, N.; Zhang, X.Y. Analysis on influence of propellant component on interior ballistics of combustion light gas gun. *J. Ballist.* **2012**, *24*, 90–93.
19. Bird, G.A. *A Note on Combustion Driven Tubes*; Royal Aircraft Establishment: Farnborough, UK, 1957; Volume 146.
20. Lu, F.K.; Wilson, D.R.; Bakos, R.J.; Erdos, J.I. Recent Advances in Detonation Techniques for High-Enthalpy Facilities. *AIAA J.* **2000**, *38*, 1676–1684. [CrossRef]
21. Presles, H.N.; Bauer, P. Detonation products gun. *Rev. Sci. Instrum.* **1983**, *54*, 1511–1512. [CrossRef]

22. Batchelor, P.L. Development of a gaseous detonation driven hypervelocity launcher. Master's Thesis, University McGill, Montreal, QC, Canada, 2010.
23. Li, J.P.; Chen, H.; Zhang, S.Z.; Zhang, X.Y.; Yu, H.R. A gasdynamic gun driven by gaseous detonation. *Rev. Sci. Instrum.* **2016**, *87*, 15112. [CrossRef] [PubMed]
24. Sichel, M.; Tonello, N.A.; Oran, E.S.; Jones, D.A. A two–step kinetics model for numerical simulation of explosions and detonations in H2-O2 mixtures. *Proc. Math. Phys. Eng. Sci.* **2002**, *458*, 49–82. [CrossRef]
25. Jiang, Z.L. Dispersion conditions for non-oscillatory shock capturing schemes and its applications. *Comput. Fluid Dyn. J.* **1995**, *2*, 137–150.
26. Li, J.P.; Jiang, Z.L.; Chen, H.; Feng, H.; Yu, H.R. Numerical study on backward-forward double-detonation driver for high enthalpy shock tubes. *Chin. J. Theor. Appl. Mech.* **2007**, *3*, 343–349.
27. Jacobs, P.A. Quasi-one-dimensional modeling of a free-piston shock tunnel. *AIAA J.* **1994**, *32*, 137–145. [CrossRef]
28. Chue, R.S.M.; Tsai, C.Y.; Bakos, R.J. Driver gas contaminationin a detonation-driven reflected-shock tunnel. *Shock Waves* **2003**, *13*, 367–380. [CrossRef]
29. Grossir, G.; Ilich, Z.; Chazot, O. Modeling of the VKI Longshot Gun Tunnel Compression Process Using a Quasi-1D Approach. In Proceedings of the 33rd AIAA Aerodynamic Measurement Technology and Ground Testing Conference, Denver, CO, USA, 5–9 June 2017; American Institute of Aeronautics and Astronautics: Reston, VA, USA, 2017; p. 3985.
30. Ebnesajjad, S. *Expanded PTFE Applications Handbook: Technology, Manufacturing and Applications*; William Andrew: Norwich, NY, USA, 2016; pp. 163–170.

MDPI
St. Alban-Anlage 66
4052 Basel
Switzerland
www.mdpi.com

Applied Sciences Editorial Office
E-mail: applsci@mdpi.com
www.mdpi.com/journal/applsci

www.ingramcontent.com/pod-product-compliance
Lightning Source LLC
Chambersburg PA
CBHW061559210926
51850CB00009B/2156
* 9 7 8 3 0 3 6 5 9 0 3 3 2 *